# CALCULUS

# THE EASY WAY

# CALCULUS
# THE EASY WAY

**Douglas Downing**

Yale University

BARRON'S EDUCATIONAL SERIES, INC.
New York • London • Toronto • Sydney

## Acknowledgments

The inspiration for this book came from the Shoreline High School 1975 calculus class and its teacher, Mr. Clint Carlson. I am also indebted to my teachers in physics, astronomy, economics, and mathematics at Yale University. My thanks go also to my mother, my sister Marlys, and Mr. Mark Yoshimi for their help in reviewing the manuscript.

*All inquiries should be addressed to:*
Barron's Educational Series, Inc.
250 Wireless Boulevard
Hauppauge, New York 11788

International Standard Book No. 0-8120-2588-1

PRINTED IN THE UNITED STATES OF AMERICA

7   510   17 16

Illustrations by Susan Detrich

# Contents

## List of Symbols

| | |
|---|---|
| $\sqrt{\phantom{x}}$ | square root |
| $\mid\mid$ | absolute value |
| $\pm$ | plus or minus |
| $<$ | less than |
| $>$ | greater than |
| $\approx$ | approximately equal to |
| $\infty$ | infinity |
| $\int$ | integral sign |

GREEK LETTERS

| | |
|---|---|
| $\Delta$ | capital delta (used for ''change in'') |
| $\Sigma$ | capital sigma (used for summation) |
| $\rho$ | rho (represents density) |
| $\pi$ | pi (= 3.14159 . . . .) |
| $\Omega$ | capital omega (represents angular frequency) |
| $\omega$ | lowercase omega (represents angular frequency) |
| $\theta$ | theta |

# Introduction

This book tells of adventures that took place in the land of Carmorra. The story is told here because, by following these adventures, you can learn differential and integral calculus. This book includes material suitable for a first-year calculus course. It is designed to be used in a classroom, but it can also be used by someone wishing to learn calculus on his or her own, or as a supplement to a course. This book is unlike regular math books, though. You are invited to read the book as you would read a fantasy novel.

The subject of calculus stands at the gateway to much of higher mathematics, and to applications in many different fields such as physics, biology, chemistry, economics, business, and statistics. In arithmetic, operations are carried out on numbers; in algebra, operations are carried out on symbols that stand for numbers; whereas, in calculus, operations are carried out on functions that represent the relationship between two variable quantities. Some integration techniques date back to the time of the ancient Greek world, but what we now know as calculus was developed independently by Isaac Newton in 1666 and Gottfried Wilhelm Leibniz in 1675. Newton called his invention the method of *fluxions,* which he developed at the same time that he was developing the foundations of the branch of physics known as mechanics.

You will best appreciate this book if you have about the same mathematical background as the people of Carmorra had at the beginning of the story. The material in this book is designed to follow high school courses in algebra, trigonometry, and geometry. You should know basic algebra terminology and methods, such as how to solve an equation with the quadratic formula. Experience in factoring second-degree polynomials will also be beneficial. Calculus depends heavily on analytic geometry, so it helps if you are familiar with Cartesian coordinates, the slopes of lines, and the equation of figures such as circles, ellipses, and parabolas. Function notation, as in $f(x) = x^2$, is also used extensively throughout the book.

You should be familiar with basic trigonometric functions and know some of their properties. A review list of trigonometric identities is included in Chapter 11. Some familiarity with logarithmic and exponential functions will help, although it is not essential to understand the book. Imaginary numbers play a small role in Chapter 15, but familiarity with imaginary numbers is not needed anywhere else in the book.

This book is designed to let you solve applied problems as quickly as possible. Many of the results presented here are demonstrations rather than formal proofs. If you are planning further study in calculus, you should become familiar with some of the rigorous background theory, such as the meaning of continuity and of limit.

The people in Carmorra use a very bizarre system of measurement, so I have translated numerical measurements into the metric system or else left measurements in terms of general units. If you are a science student, in particular, you will have to learn to be rigorous in your treatment of units.

At the end of each chapter are exercises to provide practice in applying the concepts developed in that chapter. Understanding any mathematical material requires work. The answers are provided at the back of the book so you can tell for yourself how well you have mastered the problems. The problems in Chapter 16 provide a comprehensive test of material from throughout the book. The exercises came from a wide variety of sources; some were supplied by the gremlin and some were dreamed up by Professor Stanislavsky. The final test in Chapter 16 is presented here exactly as the gremlin presented it to us.

The ready availability of pocket calculators opens new possibilities for students learning calculus. If you have a calculator available, you might like to verify many of the numerical results found in the text. Some of the exercises are obviously intended to be done with the aid of a calculator and should not be attempted without one. If you do not have a calculator available, you will have to take my word for the numerical results.

If you have never seen a calculus problem before, you are in the same position that Recordis, the professor, and the others are at the beginning of the story. You are now about to embark on the discovery of calculus.

# CALCULUS

# THE
# EASY
# WAY

# 1

# The Slope of the Tangent Line

The storm struck my ship with devastating suddenness. Something hit me on the head, and my memory was completely knocked out. The next thing I remember was being washed ashore on a strange land called Carmorra. The farmer who first met me, Mr. Floran, decided to take me to the capital city.

There it proved to be a time of crisis. Nobody was able to figure out the speed of the new train, which was powered by a friendly giant named Mongol. Mongol pushed the train with a constant force while the train kept going faster and faster, until Mongol decided to play with something else.

Mr. Floran and I rode the train from Coast City to the Capital, where he took me to the Royal Palace. A heated debate was going on in a room labeled "Main Conference Room."

"Have you made any more progress?" Mr. Floran asked after he had introduced me.

"No," a pleasant woman with intense blue eyes said sadly. ("That's Professor Stanislavsky," Floran whispered to me.)

"Yes, we have!" contradicted a middle-aged man with an elaborately carved pen in his hand and three more pens behind his ear. "It has been proved to be impossible to solve the problem." ("That's Marcus Recordis, the Royal Keeper of the Records," Floran whispered.)

"We have indeed seemingly reached an impasse," a man in a glittering robe said. ("That's the king," Floran informed me. "You had better bow to him.") After the necessary formalities were over, Floran introduced me to the other people in the room: Alexanderman Trigonometeris, the Royal Keeper of the Triangles, and Gerard Macinius Builder, the Royal Construction Engineer.

"You mustn't forget Igor," Recordis said.

"Who is Igor?" I asked, seeing no one else in the room.

"This is Igor," Recordis said, slapping a large object on the wall that looked like a combination television screen and blackboard. "This is the only Visiomatic Picture Chalkboard Machine in the world."

"Now we can explain the problem," the professor stated. "up to now we have not been able to use Mongol to his full capacity because of limitations on our frictionless track. If the speed of the train ever exceeded a certain amount, the track would break and there would be a terrible accident. The trouble is that we don't know how fast the train is going at a given time."

"Explain what you mean," I said.

"It is a simple matter to tell the position of the train at any time. Draw the picture, Igor. (See Figure 1–1.) Recordis boards the train with his watch. All along the track we have markers telling how far it is from the start at the ocean. Every time 1 minute goes by, Recordis shouts 'Now!' and Trigonometeris quickly looks outside and writes down how far the train has gone. They made a table of their results." (See Table 1–1.)

"With these numbers it is easy to make a position-time graph, like the one the professor just showed you," the king said. "Igor draws two perpendicular lines, marks time on the horizontal line and distance on the vertical line, and then puts a dot at every point where the time number di-

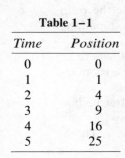

**Table 1–1**

| Time | Position |
|------|----------|
| 0 | 0 |
| 1 | 1 |
| 2 | 4 |
| 3 | 9 |
| 4 | 16 |
| 5 | 25 |

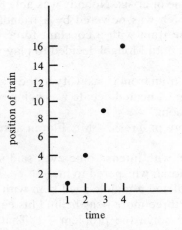

Figure 1–1.

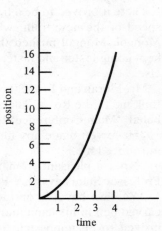

Figure 1–2.

rectly under the point is equal to the time when the train is at the position number directly to the left of it.

"The next thing we can do with the graph is figure out exactly where the train was at some time in the middle, such as 3.5 minutes. Obviously, the train must be somewhere on the track, and it must be somewhere between the place where it was at 3 minutes and the place where it was at 4 minutes (because it always goes in the same direction). The only way to figure out exactly where it is is to draw a line or a curve that connects all the points on the graph. The most logical connecting line is a smooth curve." (See Figure 1–2.)

"This way we can represent the entire curve mathematically," Recordis said. "It just so happens that the distance that the train has moved from the starting point is equal to the amount of time it has been traveling multiplied by itself."

(distance train has moved) = (time in minutes) × (time in minutes)

"We can abbreviate this equation by denoting the distance traveled by some letter, such as $d$."

"Why?" the king asked.

"All right, call it $y$ if you prefer," Recordis said. "Then I suppose we should call the time the train has been traveling $x$."

$$y = (x) \times (x)$$

"That is the same thing as writing $x^2$," the king pointed out.

$$y = x^2$$

"We can also write that as a function machine," the professor said. "We decided that a *function* was a machine that turned one number into

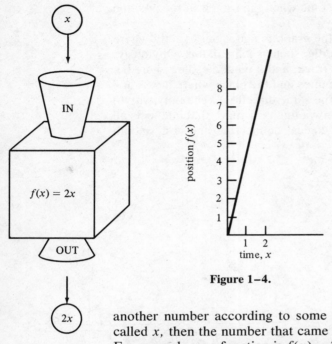

Figure 1–3.

Figure 1–4.

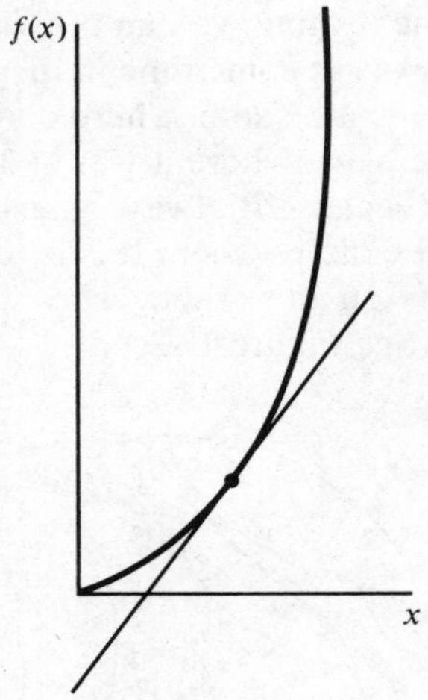

Figure 1–5.

another number according to some rule. If the number we put in was called $x$, then the number that came out of the machine was called $f(x)$. For example, one function is $f(x) = 2x$." (Figure 1–3.)

"In this case it is easy to figure out what number will come out," Recordis said. "If we put in 2, we'll get 4; if we put in 10, we'll get 20; et cetera."

"Also, we can call the in-number anything we like. We don't have to call it $x$," the king added. "If $f(x) = 2x$, then we can also say that $f(q) = 2q$, $f(a + b) = 2(a + b)$, and $f(3x^2 + 4x + 5) = 6x^2 + 8x + 10$. We can say that $f((\text{in-number})) = 2 \times (\text{in-number})$."

"I remember when we made lots of functions," Recordis said. "We had $f(x) = x^2$, $f(x) = 3x^3 + 2x^2 + x + 4$, $f(x) = \sin x$, and lots of others."

"In this case we want to use the first one you mentioned," the professor stated. "We want $f(x) = x^2$. In this case $x = $ time and $f(x) = $ distance that the train has traveled. We don't have any trouble getting this far. We can tell the position of the train at any given time, but we need to know the speed."

"We almost know what it is," the king said. "We know that speed is given by the slope of the curve. For example, if Recordis walks at a constant speed of 4 units per minute, then his position function is given by $f(x) = 4x$. If you make a graph of his position, it looks like a straight line." (Figure 1–4.)

"How can you figure out the slope of a curve, like $f(x) = x^2$?" I asked. "At one point the curve is sloped very gently, but at other points it is sloped very steeply."

"Right," the professor said. "The slope of the curve is changing because the speed of the train is changing. If the speed of something is changing, we can define only its instantaneous speed, that is, the speed it is traveling at a given instant. This means that we need to calculate an instantaneous slope for the curve. You were right when you said that we can't determine the slope of a curve. But we can draw a line right next to the curve that has the same slope as the curve does at that point." (Figure 1–5.)

"We call that line the *tangent line* for the curve," the king told us. "Notice that this curve has lots of different tangent lines." (Figure 1–6.)

"You should also notice that the tangent lines intersect the curve at one and only one point," Recordis said.

"That means that we can define the slope of the curve at a given point to be equal to the slope of the tangent line at that point," the professor added.

"Therefore, to find the speed of the train, all we need to do is find the slope of the tangent line to the curve," the king noted.

"That is what we have proved to be impossible," Recordis said heatedly. "We know very well how to draw a line. First, we start out with two points. We can call them anything we like, say $(a, b)$ and $(c, d)$. Now we can easily compute the slope of the line between these two points. A long time ago we defined the slope as equal to the distance the line goes up divided by the distance the line goes sideways. (See Figure 1–7.)

$$(\text{slope}) = \frac{(\text{up})}{(\text{sideways})}$$

"We know that (up) is equal to $(d - b)$, and that (sideways) is equal to $(c - a)$. This lets us say that the slope of the line is equal to $(\text{slope}) = (d - b)/(c - a)$. This method works for any line in the world, as long as we

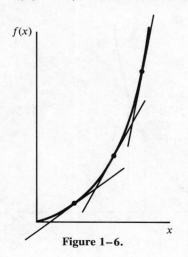

Figure 1–6.

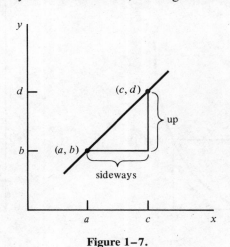

Figure 1–7.

know *two* points on it. There is absolutely no way to find the slope of the tangent line, though, because we know only one point! We know lots of other points that are not on the line, but we don't know one other single point that *is* on the line.''

There was a long silence as we contemplated what he had said.

''We must find the answer to this problem,'' the king stated. ''I don't care how we have to do it.''

''There has to be some solution,'' the professor said. ''In fact, we placed a large wager with the gremlin that we would be able to reach a solution in the next few days.''

''Who is the gremlin?'' I asked.

''He is our arch-enemy,'' the king told me. ''It is his sole purpose to disrupt our entire learning process and take over the kingdom of Carmorra. We have already defeated him several times concerning matters of algebra.''

I looked at the drawing of the graph for several minutes. Finally I said, ''It appears that our problem is that we need to find another point someplace. Recordis is certainly right when he says that we need two points to determine the slope of the line. Igor, draw a graph showing the point where we want to find the slope of the tangent line. (See Figure 1–8.)

''Let's say that this point represents the train at some time called *a*. This means that the distance the train has traveled equals $a^2$. The coordinates of this point are $(a, a^2)$. We still have our position-locating function machine, so we can also write the coordinates as $(a, f(a))$. In order to find the slope of the tangent line, we need another point. Since the only points that we know very much about are the other points on the curve, we will have to use one of those. Igor, show me another point on the curve that is close to the first point. (See Figure 1–9.)

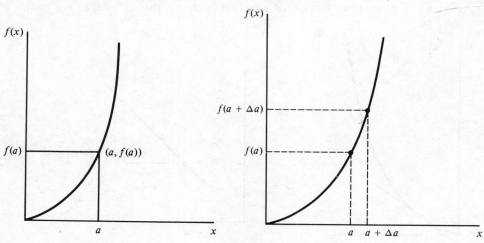

Figure 1–8.                    Figure 1–9.

"It really doesn't make much difference how far away the second point is from the first point, so we can make up some distance and call it $\Delta a$. (The little triangle $\Delta$ is the fourth letter—capital form—of the Greek alphabet. It is known as *delta*. The symbol $\Delta a$ is pronounced 'delta-a.') Then we know that the $x$ coordinate, or the time coordinate, of the second point is equal to $(a + \Delta a)$."

"The $y$ coordinate is still $y = f(x)$, so we can plug that into the machine and say that the $y$ coordinate is equal to $f(a + \Delta a)$," the professor noted.

"I know the slope of the line between those two points," Recordis said.

$$\text{(slope of line between these two points)} = \frac{f(a + \Delta a) - f(a)}{a + \Delta a - a}$$

$$= \frac{f(a + \Delta a) - f(a)}{\Delta a}$$

"That line still isn't very close to the tangent line, though," Recordis added.

"Maybe we can make the line between these two points move closer to the tangent line," I said. "Igor, make a sketch of where you think the tangent line should be." (Figure 1–10.)

"We can call the top line a *secant* line," the professor stated. "We use that name for a line that intersects a curve in two points, rather than one point as the tangent line does."

"All right," I said. "This means that we need to somehow make the secant line, whose slope we do know, move closer to the tangent line, whose slope we don't know."

The king looked closely at the picture. "Couldn't we make the second

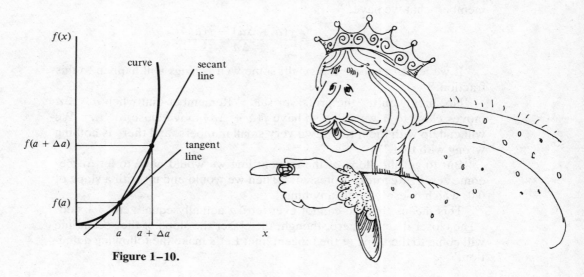

Figure 1–10.

point move closer to the first point? Wouldn't that make the secant line approach the tangent line?''

''That's it!'' I said. ''We can make the two points move closer together by making $\Delta a$ smaller and smaller. Igor, draw a series of pictures showing what happens when the two points move closer together.'' (Figure 1–11.)

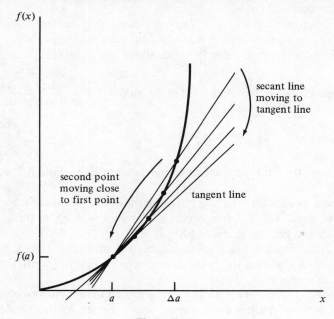

**Figure 1–11.**

''What happens to the expression for the slope?'' Recordis asked. ''Remember what we have.''

$$(\text{slope}) = \frac{f(a + \Delta a) - f(a)}{\Delta a}$$

''If we let $\Delta a$ become too small, some weird things will happen to this fraction.''

''No, they won't,'' the professor said. ''Remember that when $a + \Delta a$ moves close to $a$ we will also have $f(a + \Delta a)$ move close to $f(a)$. We will end up with the ratio of two very small numbers, and there is nothing wrong with that.''

''But to get the slope of the tangent line we would have to let $\Delta a$ become zero!'' Recordis protested. ''Then we would end up with a slope of 0/0, which doesn't tell us anything.''

''This means that we cannot ever let $\Delta a$ actually equal zero,'' I said. ''The closer it gets to zero, though, the closer the slope of the secant line will come to the slope of the tangent line. Let's make the following definition.''

$$(\text{slope of tangent line}) = \lim_{\Delta a \to 0} \frac{f(a + \Delta a) - f(a)}{\Delta a}$$

"What does that mean?" Recordis protested. "What do you mean by that 'limit' thing?"

"I understand," the professor said. "We will let $\Delta a$ move very, very, very, very close to zero, but we will put up a little fence that prevents it from ever actually equaling zero."

"That's all very nice theoretically," the king remarked. "It still doesn't tell us how to find a number that represents the slope of the tangent line, though."

"We know what $f(x)$ is," I said. "We can rewrite the equation, only this time we will use $f(x) = x^2$."

$$(\text{slope}) = \lim_{\Delta a \to 0} \frac{(a + \Delta a)^2 - a^2}{\Delta a}$$

"I know what $(a + \Delta a)^2$ is," Recordis said. "That's algebra."

$$(\text{slope}) = \lim_{\Delta a \to 0} \frac{a^2 + 2a\,\Delta a + \Delta a^2 - a^2}{\Delta a}$$

"The $a^2$ and the $-a^2$ will cancel out," the king pointed out helpfully.

$$(\text{slope}) = \lim_{\Delta a \to 0} \frac{2a\,\Delta a + \Delta a^2}{\Delta a}$$

"We can divide both the top and the bottom of the fraction by $\Delta a$," the professor said. "After all, we never let $\Delta a$ become zero."

$$(\text{slope}) = \lim_{\Delta a \to 0} \frac{(1/\Delta a)(2a\,\Delta a + \Delta a^2)}{(1/\Delta a)(\Delta a)}$$

$$= \lim_{\Delta a \to 0} \frac{2a + \Delta a}{1}$$

"Now it is very clear," I said. "As $\Delta a$ approaches zero, the quantity $(2a + \Delta a)$ will approach $2a$."

"That's the answer!" the king shouted.

"That is the slope of the tangent line," the professor said reverently.

"That's amazing!" Trigonometeris said.

"That's too simple," Recordis protested suspiciously.

"We should make sure that it makes sense," I said cautiously. "Let's look at the graph. At the place where $a = 0$, we know that the train has not yet started to move, so its speed must be zero. According to our formula for the slope, the slope should equal $2a$, which is $2 \cdot 0$, which is zero, so it looks right."

"And if we draw the tangent line at the point where $a = 0$, it is parallel to the axis," the professor added. "Our formula should work for any function, so we should record this result as our first definition."

$$(\text{slope of tangent line}) = \lim_{\Delta x \to 0} \frac{f(x + \Delta x) - f(x)}{\Delta x}$$

"We have to think of a name for the subject we are getting into now," Recordis said. "We must do this systematically. I think I will have to start a new page in my record book."

Everybody thought of a name, but nobody came up with one that was satisfactory to all. The main problem was jealousy. Each person in the room wanted the subject named after him- or herself. The others finally turned to me and asked me to make up a name. I had vague memories of doing this sort of problem before, although I could not remember any details. For some reason the word "calculus" popped into my head, so I suggested that we call the subject *calculus*. Everyone agreed to this suggestion because the name sounded impressive.

"We will have to come back to this tomorrow," the professor said. "We can try other functions and see how they work out. First we should record what we found today."

$$\text{slope of tangent line for } f(x) = x^2 \text{ is } 2x$$

The group adjourned amidst great excitement, and they ran to the train to tell everyone that they knew how fast it went. Farmer Floran decided that he had better return to his home, so he boarded the train and Mongol pushed him back to Coast City.

Later in the evening it was decided that in gratitude for my services I would be provided with lodging in the palace. I told the king that I wanted to go home as soon as I remembered where I had came from, but that I would be glad to stay and help for a while. In turn, the king promised to

help me when I was ready to return home. "I think you arrived at the beginning of an exciting period," the king told me. "I wonder what we will
discover tomorrow."

## Exercises

1. The following pairs of points all define secant lines to the curve $y = x^2$ through the point (2, 4). Find the slope of each secant line: (2, 4) and (3, 9); (2, 4) and (2.5, 6.25); (2, 4) and (2.3, 5.29); (2, 4) and (2.1, 4.41); (2, 4) and (2.05, 4.20).

2. Find the slope of the secant line defined by each of these pairs of points: (1, 1) and (2, 4); (1.5, 2.25) and (2, 4); (1.7, 2.89) and (2, 4); (1.8, 3.24) and (2, 4); (1.9, 3.61) and (2, 4); (1.95, 3.80) and (2, 4).

3. Find the equation of the tangent line to the curve $y = x^2$ at the point (2, 4).

4. Find the equation of the tangent line to the curve $y = x^2$ through the point (7, 49). Use this tangent line to estimate $\sqrt{50}$.

5. Draw a graph of the function

$$y = f(x) = \frac{8x + x^2}{x}$$

What is $f(0)$? What is $\lim\limits_{x \to 0} f(x)$?

# 2
# Calculating Derivatives

Everybody gathered around Igor the next morning. The professor said that she had a whole series of new ideas to try out. Recordis started the meeting with an anguished complaint.

"This will never do!" he cried. "We must think of a shorter name for this whatever-it-is we've discovered. I can't write 'slope of the tangent line' all day. Already my wrist is developing a terrible cramp."

"Then we shall think of a name," the professor said matter-of-factly. "Let's look at our definition again."

$$\text{(slope of tangent line)} = \lim_{\Delta x \to 0} \frac{f(x + \Delta x) - f(x)}{\Delta x}$$

(We had decided to write "lim" as an abbreviation for "limit.")

"Now, what does that look like?" the professor asked.

"It looks to me like the time Mongol spilled his letter blocks and we never could figure out all the words he had made," Trigonometeris said.

"We must take this seriously," the king rebuked. "We must think of a real name."

Everybody suggested names, but nothing sounded satisfactory. Finally the professor turned to me and said, "You were able to think of the name 'calculus.' Maybe you can think of a name for the slope of the tangent line."

12

I thought for a couple of minutes and came up with another name. "We could call it a *derivative*," I suggested.

"That sounds as good as any," Recordis said. "It surely was frustrating to derive it."

function $y = f(x)$

derivative = slope of tangent line = $\lim_{\Delta x \to 0} \dfrac{f(x + \Delta x) - f(x)}{\Delta x}$

"We still need to think of a symbol to stand for derivative," Recordis complained. "I can't write the word 'derivative' all the time."

Everyone looked at the board for a few minutes. "I have an ingenious idea," the professor said as modestly as she could. "Since a derivative is a slope, it should have units of $y/x$. For example, we wrote delta $y$ over delta $x$ ($\Delta y/\Delta x$) to stand for a tiny increment of $y$ divided by a tiny increment of $x$. Why don't we say that $dy/dx$ is the slope of the tangent line?"

Recordis was reluctant to agree to any symbol that required him to write four letters.

"I always liked the little prime (') symbol," the king said. "I liked it when we wrote $y'$ and called it 'why prime?' We could call the derivative $y'$."

The professor looked hurt, but Recordis was happy. "I really like that!" he said.

"But you could get confused!" the professor protested. "What if the variable in the function isn't $x$? What if you have $y = f(t)$, $y = f(w)$, or $y = f(q)$? In my system you could write $dy/dt, dy/dw,$ or $dy/dq$."

"But look at all that writing!" Recordis said.

"I don't think we need to have an argument here," I told them. "We'll use both systems. At any particular time we'll use whichever one seems to be the more convenient."

function $y = f(x)$

derivative $y' = f'(x) = \dfrac{dy}{dx} = \lim_{\Delta x \to 0} \dfrac{f(x + \Delta x) - f(x)}{\Delta x}$

"We wasted too much time thinking of names," the professor said. "We must start with the important part. We must be systematic about this, and make a list of different kinds of functions and their derivatives."

"It seems to me that the simplest function is one that has the same value all the time," the king stated.

"I remember when we made an $f(x) = 2$ function," Recordis said. "Mongol got tired of 2's, but no matter what number he put into the function machine he always got a 2 out."

"Let's see what our formula says if $y = 2$," the professor said.

$$f(x) = 2$$

$$f'(x) = \lim_{\Delta x \to 0} \frac{f(x + \Delta x) - f(x)}{\Delta x}$$

$$= \lim_{\Delta x \to 0} \frac{2 - 2}{\Delta x}$$

$$= \lim_{\Delta x \to 0} \frac{0}{\Delta x}$$

$$f'(x) = 0$$

"We should have a slope of zero," the professor said.

"Let's draw a graph of the function $y = 2$, just to make sure," Recordis said. (See Figure 2–1.)

"That's just a straight line with no slope," Trigonometeris pointed out. "I could have told you the slope of that before we developed all this hocus-pocus."

"But we know that zero must be the right answer," the king said. "Let's pretend that the graph represents the position of the train. Then this means that the train is 2 units from the ocean and is just staying in one place. If it is stopped, then of course its speed is zero."

"Like the time Mongol was scared by that parakeet and just stopped in the middle of the track, still holding onto the train," Recordis reminded us.

"The same thing should happen if $y$ is any constant number, and not just 2," the professor said. "We could have $y$ equals 2 or 7.6 or $819\frac{1}{2}$ or anything else, just so long as it doesn't change. We can write that as our next rule: The derivative of a constant function is zero."

| Function | Derivative |
|----------|------------|
| $y = c$ | $y' = dy/dx = 0$    (when $c$ is a constant) |

**Table 2–1**

| $x$ | $y$ |
|-----|-----|
| $-1$ | $-2$ |
| 0 | 0 |
| 1 | 2 |
| 2 | 4 |
| 3 | 6 |

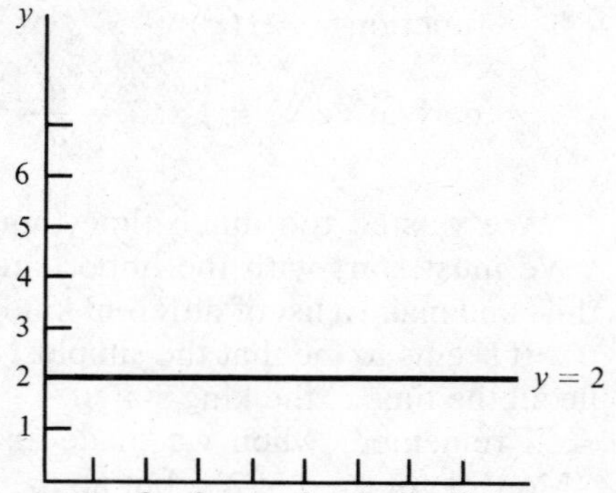

**Figure 2–1.**

"What if we have a tilted line?" Trigonometeris asked. (Figure 2–2.)

"We've had functions like this before. Igor, show us a table of values." (Table 2–1.)

"That function is easy to recognize," the professor said. "We have $f(x) = 2x$. Let's plug that into our formula."

$$\frac{dy}{dx} = y' = \lim_{\Delta x \to 0} \frac{2(x + \Delta x) - 2(x)}{\Delta x}$$

$$= \lim_{\Delta x \to 0} \frac{2x + 2\Delta x - 2x}{\Delta x}$$

$$= \lim_{\Delta x \to 0} \frac{2\Delta x}{\Delta x}$$

$$\frac{dy}{dx} = 2$$

"But I could have told you the slope was 2," Trigonometeris said. "We could have figured out the slope using the old method." (Figure 2–3.)

$$(\text{slope}) = \frac{(\text{up})}{(\text{sideways})} = 2$$

"I don't think you really need calculus methods whenever you have any function that is a straight line," the professor stated. "However, it is a good thing that the calculus methods give us the same answer for the slope as the regular methods did. If calculus turned out to be inconsistent with algebra and geometry, we'd be in real trouble."

"I bet that we can generalize this rule," the king said. "Suppose that $f(x)$ equals $x$ times any constant number, say $c$."

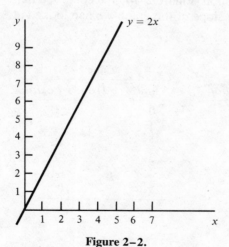

Figure 2–2.

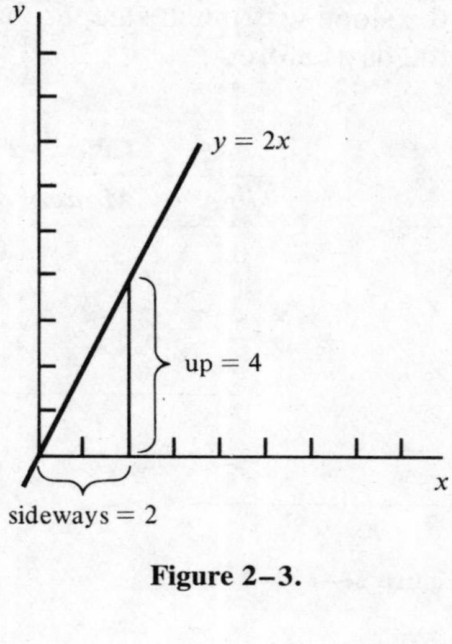

Figure 2–3.

$$f(x) = cx$$

$$f'(x) = \lim_{\Delta x \to 0} \frac{c(x + \Delta x) - cx}{\Delta x}$$

$$= \lim_{\Delta x \to 0} \frac{cx + c\,\Delta x - cx}{\Delta x}$$

$$= \lim_{\Delta x \to 0} \frac{c\,\Delta x}{\Delta x}$$

$$f'(x) = c$$

"That looks like a good rule," the professor agreed.

| Function | Derivative |
|---|---|
| $y = cx$ | $y' = dy/dx = c$    (when $c$ is a constant) |

"I know something that has a position function that looks like that," Recordis said. "Remember when we took Mongol to Ice Skating Lake and gave him a push? We made a table of his position at different times." (Table 2–2.)

"What happened after that?" I asked.

"The ice cracked and he almost fell through," the professor answered. "I remember that we established that he had been traveling with a constant speed of 3 units per second. Apparently whenever something travels with a constant speed the derivative of its position function is a constant number."

"I remember a complicated situation," Recordis said. "Remember the time we fed Mongol some Extra-Strength Tablets before we let him play with the train? He started running very fast, and we kept a table of values." (Table 2–3.)

Igor drew a graph of the train's motion (Figure 2–4). We could see that the slope was much steeper than the slope of the graph we had looked at the day before.

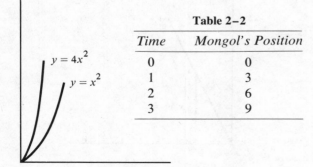

**Table 2–2**

| Time | Mongol's Position |
|---|---|
| 0 | 0 |
| 1 | 3 |
| 2 | 6 |
| 3 | 9 |

**Table 2–3**

| Time | Train's Position |
|---|---|
| 0 | 0 |
| 1 | 4 |
| 2 | 16 |
| 3 | 36 |
| 4 | 64 |
| 5 | 100 |

Figure 2–4.

"I recognize that function," the king said. "One squared times 4 is 4, 2 squared times 4 is 16, and 3 squared times 4 is 36."

"That's it," the professor said. "The function is $f(x) = 4x^2$."

We put that in the formula for the derivative:

$$f(x) = 4x^2$$

$$f'(x) = \lim_{\Delta x \to 0} \frac{4(x + \Delta x)^2 - 4x^2}{\Delta x}$$

$$= \lim_{\Delta x \to 0} \frac{4(x^2 + 2x\,\Delta x + \Delta x^2) - 4x^2}{\Delta x}$$

$$= \lim_{\Delta x \to 0} \frac{4x^2 + 8x\,\Delta x + 4\,\Delta x^2 - 4x^2}{\Delta x}$$

$$= \lim_{\Delta x \to 0} \frac{8x\,\Delta x + 4\,\Delta x^2}{\Delta x}$$

$$= \lim_{\Delta x \to 0} \frac{(8x + 4\,\Delta x)\,\Delta x}{\Delta x}$$

$$= \lim_{\Delta x \to 0} 8x + 4\,\Delta x$$

$$f'(x) = 8x$$

"That means the train's speed is $8x$!" the professor said. "It was going four times faster than usual."

"In other words, 5 minutes after the train left the station it was doing $5 \times 8$, or 40, units per minute," Recordis added.

"Maybe we can generalize this rule to see what happens when $f(x) = cx^2$, where $c$ is any constant number," the professor said.

$$f(x) = cx^2$$

$$f'(x) = \lim_{\Delta x \to 0} \frac{c(x + \Delta x)^2 - cx^2}{\Delta x}$$

$$= \lim_{\Delta x \to 0} \frac{cx^2 + 2cx\,\Delta x + c\,\Delta x^2 - cx^2}{\Delta x}$$

$$= \lim_{\Delta x \to 0} \frac{2cx\,\Delta x + c\,\Delta x^2}{\Delta x}$$

$$f'(x) = 2cx$$

"That works when $c = 1$, because then we get $f'(x) = 2x$, which is what we did yesterday," the professor reminded us.

"Or it works when $c = 4$, because then we get $f'(x) = 8x$, which is what we just did," Recordis said.

"Or it even works when $c = 0$," the king noted. "Then we get $f(x) = 0$, which is a constant number so $f'(x) = 0$."

We added this new rule to our list:

Function                 Derivative
$y = cx^2$               $y' = dy/dx = 2cx$

"I remember once when the train didn't start from the ocean," Recordis said. "Mongol started pushing when we were 5 units away from the zero point. We made a table of values." (Table 2–4.)

Igor drew a graph (Figure 2–5).

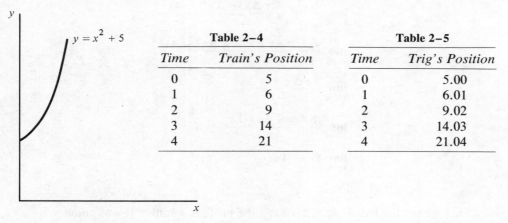

Table 2–4

| Time | Train's Position |
|------|------------------|
| 0 | 5 |
| 1 | 6 |
| 2 | 9 |
| 3 | 14 |
| 4 | 21 |

Table 2–5

| Time | Trig's Position |
|------|-----------------|
| 0 | 5.00 |
| 1 | 6.01 |
| 2 | 9.02 |
| 3 | 14.03 |
| 4 | 21.04 |

**Figure 2–5.**

"I think the function is $f(x) = x^2 + 5$," the king suggested. "Let's try plugging that function into our formula for the derivative."

$$f(x) = x^2 + 5$$

$$f'(x) = \lim_{\Delta x \to 0} \frac{(x + \Delta x)^2 + 5 - (x^2 + 5)}{\Delta x}$$

$$= \lim_{\Delta x \to 0} \frac{x^2 + 2x\,\Delta x + \Delta x^2 + 5 - x^2 - 5}{\Delta x}$$

"We can cancel out the $(x^2 + 5)$ and the $(-x^2 - 5)$," Recordis noted.

$$f'(x) = \lim_{\Delta x \to 0} \frac{2x\,\Delta x + \Delta x^2}{\Delta x}$$

$$= \lim_{\Delta x \to 0} 2x + \Delta x$$

$$= 2x$$

"That's the same answer we got when Mongol started pushing from point zero," Recordis said.

"It seems as though it should be the same answer," the king remarked.

"Mongol pushes the same amount each time, so it seems as though the speed of the train at a given instant should not depend on where he started pushing."

"I remember one time that was very complicated," Recordis said. "The train started 5 units away from the ocean, and Trigonometeris started running inside the train. We made a table of his position." (Table 2–5.)

"I was very careful to make sure that I ran at a constant speed with respect to the train," Trigonometeris stated proudly.

"I know what position function we need to use," the king said. "It will be exactly the same as the time before, except this time we must add the distance that Trig has walked from the back of the train. Let's try the following function."

$$(\text{Trig's position at time } x) = f(x) = x^2 + (0.01)x + 5$$

We put that function into the formula for the derivative.

$$f(x) = x^2 + (0.01)x + 5$$

$$f'(x) = \lim_{\Delta x \to 0} \frac{(x + \Delta x)^2 + (0.01)(x + \Delta x) + 5 - x^2 - (0.01)x - 5}{\Delta x}$$

$$= \lim_{\Delta x \to 0} \frac{x^2 + 2x\,\Delta x + \Delta x^2 + (0.01)x + (0.01)\,\Delta x - x^2 - (0.01)x}{\Delta x}$$

$$= \lim_{\Delta x \to 0} \frac{2x\,\Delta x + \Delta x^2 + (0.01)\,\Delta x}{\Delta x}$$

$$= \lim_{\Delta x \to 0} 2x + \Delta x + 0.01$$

$$f'(x) = 2x + 0.01$$

While we were admiring this answer, the king said, "I just noticed something: $2x$ is the speed of the train, and 0.01 is the speed of Trig as he walks along inside the train. It looks as though you just add them together."

"Fascinating," the professor stated. "The original function was $f(x) = x^2 + (0.01)x + 5$. The first term represents the position of the train, the second term represents the position of Trig with respect to the train, and the third term is a constant which, of course, has a derivative of zero. Maybe, if you have a sum of functions, you can take the derivative of each term and add them together to get the derivative of the whole function."

"Let's see whether we can prove that in general," the king said. "Suppose we have any two functions, say $f(x)$ and $g(x)$, and we make a new function—call it $q(x)$—which equals $f(x) + g(x)$. Let's plug that into the formula and see whether we can find the derivative of $q(x)$."

$$q(x) = f(x) + g(x)$$

$$q'(x) = \frac{dq}{dx} = \lim_{\Delta x \to 0} \frac{f(x + \Delta x) + g(x + \Delta x) - f(x) - g(x)}{\Delta x}$$

"Now we're stuck," Recordis mourned.

"I think we can rearrange these terms," the king said.

$$q'(x) = \lim_{\Delta x \to 0} \frac{f(x + \Delta x) - f(x) + g(x + \Delta x) - g(x)}{\Delta x}$$

$$= \lim_{\Delta x \to 0} \frac{f(x + \Delta x) - f(x)}{\Delta x} + \lim_{\Delta x \to 0} \frac{g(x + \Delta x) - g(x)}{\Delta x}$$

$$= f'(x) + g'(x)$$

"It does work!" The professor said in amazement. "This rule will make life much simpler. This means that, whenever we have a function made up of a whole glob of little functions added together, we can take the derivative of each little function and add all the derivatives together to get the derivative of the whole glob."

---

## SUM RULE FOR DERIVATIVES

| Function | Derivative |
|---|---|
| $q(x) = f(x) + g(x)$ | $q'(x) = f'(x) + g'(x)$ |

---

"We could write the same rule if we had three functions added together," Recordis added.

$$q(x) = f(x) + g(x) + h(x)$$

$$q'(x) = f'(x) + g'(x) + h'(x)$$

"Or even four functions added together," Recordis continued, getting carried away.

$$q(x) = f(x) + g(x) + h(x) + i(x)$$

$$q'(x) = f'(x) + g'(x) + h'(x) + i'(x)$$

"Yes, we know what you mean," the professor said quickly, before Recordis had a chance to say that they could write the same rule for five functions added together. "I think we should go on to something else."

"I remember a long time ago when we made a list of crazy functions," Recordis remarked. "I wonder if this calculus jazz will help us with any of these. Here's a good function: $f(x) = xxx$, or $f(x) = x^3$." (Figure 2–6.)

"We can try to find the derivative," the professor said.

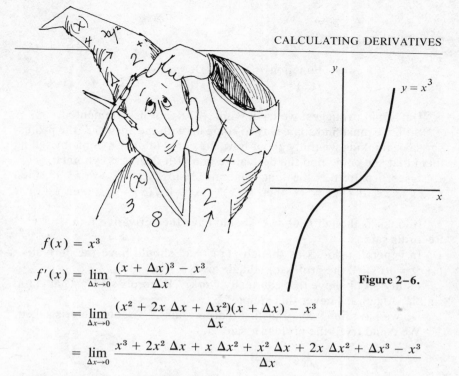

Figure 2–6.

$$f(x) = x^3$$

$$f'(x) = \lim_{\Delta x \to 0} \frac{(x + \Delta x)^3 - x^3}{\Delta x}$$

$$= \lim_{\Delta x \to 0} \frac{(x^2 + 2x\,\Delta x + \Delta x^2)(x + \Delta x) - x^3}{\Delta x}$$

$$= \lim_{\Delta x \to 0} \frac{x^3 + 2x^2\,\Delta x + x\,\Delta x^2 + x^2\,\Delta x + 2x\,\Delta x^2 + \Delta x^3 - x^3}{\Delta x}$$

The $x^3$'s canceled each other out:

$$f'(x) = \lim_{\Delta x \to 0} \frac{3x^2\,\Delta x + 3x\,\Delta x^2 + \Delta x^3}{\Delta x}$$

We factored out the $\Delta x$, which Recordis then gleefully canceled with the $\Delta x$ in the denominator. (Recordis likes to cancel things.)

$$f'(x) = \lim_{\Delta x \to 0} 3x^2 + 3x\,\Delta x + \Delta x^2$$

The last two terms went to zero when we took the limit:

$$f'(x) = 3x^2$$

"Fascinating," the king said.

"It makes sense when you look at the graph," the professor told him. (Figure 2–6.)

"When $x = 0$, we're saying that the slope of the curve is zero, which is the way it looks in the picture. As $x$ gets bigger, the slope becomes steeper. And even if $x$ is negative, the slope is still positive because $x^2$ is positive."

"How about $x^4$?" Recordis asked, wondering how complicated the world could get.

Igor slowly went through the algebra, and our eyes got tired as more and more symbols kept floating across his picture tube. "There's got to be a simpler way!" the professor said. When all of the algebra was finished, the final answer looked like this:

| Function | Derivative |
|----------|------------|
| $f(x) = x^4$ | $f'(x) = 4x^3$ |

"I'm afraid to suggest we try it with $x^5$," Recordis commented.

"Still, we must find some way to come up with the answer," the professor said, thinking wistfully about how she could impress people by telling them that she could find the derivative of a fifth-degree polynomial.

"I see a pattern," the king announced slowly. "If $f(x) = x^2$, then $f'(x) = 2x$. If $f(x) = x^3$, then $f'(x) = 3x^2$. If $f(x) = x^4$, then $f'(x) = 4x^3$."

"It looks as though $f(x) = x^5$ should have the derivative $f'(x) = 5x^4$," Recordis said.

"In general, it looks as though $f(x) = x^n$ should have the derivative $f'(x) = nx^{n-1}$," the professor added slowly.

"I still don't believe that can really work," Recordis said. "That is too simple. Algebra is never that simple."

"Do we have any way of testing that formula?" Trigonometeris asked.

"We could try," the professor said.

$$f(x) = cx^n$$

$$f'(x) = \lim_{\Delta x \to 0} \frac{c(x + \Delta x)^n - cx^n}{\Delta x}$$

That expression looked pretty hopeless, until I began to remember something. "Did you ever develop a formula for figuring out an expression like $(a + b)^n$?" I asked.

"I remember something like that," the king answered. "We derived it a long time ago when we were working out algebra. Look it up, Recordis."

Recordis fumbled through his giant book. "This might be useful," he said. "It's something called a *binomial formula*."

$$(a + b)^n = \frac{n!}{0!n!} a^n + \frac{n!}{1!(n-1)!} a^{n-1}b + \frac{n!}{2!(n-2)!} a^{n-2}b^2 + \cdots$$
$$+ \frac{n!}{(n-1)!1!} ab^{n-1} + \frac{n!}{n!0!} b^n$$

"I can't remember why it works."

"We don't need to know that now," the professor said. "The important thing is that we derived it once and that we know it does work."

"I remember the numbers with the exclamation marks," Trigonometeris stated. "They look like such excited numbers."

"That's not what that means!" the king said. "The exclamation mark is just a symbol. Remember that '5!' means 'five factorial,' and that means $5! = 5 \times 4 \times 3 \times 2 \times 1 = 120$. And, just the same, eight factorial means $8! = 8 \times 7 \times 6 \times 5 \times 4 \times 3 \times 2 \times 1 = 40{,}320$."

We put this result into the formula for the derivative.

$$f'(x) = \lim_{\Delta x \to 0} \frac{c(x + \Delta x)^n - cx^n}{\Delta x}$$

$$= \lim_{\Delta x \to 0} c \left[ \frac{n!}{0!n!} x^n + \frac{n!}{1!(n-1)!} x^{n-1} \Delta x \right.$$

$$+ \frac{n!}{2!(n-2)!} x^{n-2} \Delta x^2 + \cdots + \frac{n!}{(n-1)!1!} x \Delta x^{n-1}$$

$$\left. + \frac{n!}{n!0!} \Delta x^n \right] \frac{1}{\Delta x} - \frac{cx^n}{\Delta x}$$

"I remember when we said that $0!$ was equal to 1," the king said. "So $n!/(0!n!)$ is the same as $n!/n!$, which is equal to 1."

$$f'(x) = \lim_{\Delta x \to 0} c \left[ x^n + \frac{n!}{1!(n-1)!} x^{n-1} \Delta x + \frac{n!}{2!(n-2)!} x^{n-2} \Delta x^2 + \cdots \right.$$

$$\left. + \frac{n!}{(n-1)!1!} x \Delta x^{n-1} + \Delta x^n \right] \frac{1}{\Delta x} - \frac{cx^n}{\Delta x}$$

"I think we can simplify $n!/[1!(n-1)!]$," the professor said. "We know that $1!$ is equal to 1, so that leaves us with $n!/(n-1)!$. We can rewrite that."

$$\frac{n!}{(n-1)!} = \frac{n(n-1)(n-2)(n-3)(n-4) \ldots (4)(3)(2)(1)}{(n-1)(n-2)(n-3)(n-4) \ldots (4)(3)(2)(1)}$$

"That's great!" Recordis told her. "We can cancel out all of those, almost. The answer is that $n!/(n-1)!$ is equal to $n$."

"That simplifies our expression for the derivative a little bit," the professor said.

$$f'(x) = \lim_{\Delta x \to 0} c \left[ x^n + nx^{n-1} \Delta x + \frac{n!}{2!(n-2)!} x^{n-2} \Delta x^2 + \cdots \right.$$

$$\left. + nx \Delta x^{n-1} + \Delta x^n \right] \frac{1}{\Delta x} - \frac{cx^n}{\Delta x}$$

We multiplied through by the $c$ outside the bracket:

$$f'(x) = \lim_{\Delta x \to 0} \left[ cx^n + cnx^{n-1} \Delta x + c \frac{n!}{2!(n-2)!} x^{n-2} \Delta x^2 + \cdots \right.$$

$$\left. + cnx \Delta x^{n-1} + c \Delta x^n - cx^n \right] / \Delta x$$

Recordis happily canceled the $cx^n$ with the $-cx^n$:

$$f'(x) = \lim_{\Delta x \to 0} \left[ cnx^{n-1} \Delta x + c \frac{n!}{2!(n-2)!} x^{n-1} \Delta x^2 + \cdots \right.$$

$$\left. + cnx \Delta x^{n-1} + c \Delta x^n \right] / \Delta x$$

and the professor noted that Recordis could factor out a $\Delta x$ from the numerator:

$$f'(x) = \lim_{\Delta x \to 0} \left[ cnx^{n-1} + c\,\frac{n!}{2!(n-2)!}\,x^{n-2}\,\Delta x + \cdots + cnx\,\Delta x^{n-2} \right.$$
$$\left. + c\,\Delta x^{n-1} \right] \frac{\Delta x}{\Delta x}$$

$$= \lim_{\Delta x \to 0} cnx^{n-1} + c\,\frac{n!}{2!(n-2)!}\,x^{n-2}\,\Delta x + \cdots + cnx\,\Delta x^{n-2}$$
$$+ c\,\Delta x^{n-1}$$

"When we take the limit, $\Delta x$ will go to zero, so we will wipe out all the terms in that sum except the first one," the professor said.

$$f'(x) = cnx^{n-1}$$

"You were right!" the professor exclaimed jubilantly to the king. "That is a nice, simple formula."

| Function | Derivative |
|----------|------------|
| $f(x) = cx^n$ | $f'(x) = cnx^{n-1}$ |

"Amazing!" Trigonometeris said. "We already established that it works if $n = 2$ or $n = 3$ or $n = 4$."

"It also works if $n = 1$," the professor stated. "Then we would have $f(x) = cx$, and we already know that in that case the derivative is just equal to $c$."

"It also works if $n = 0$," the king realized. "We know that $x^0 = 1$, so then we would have $f(x) = cx^0 = c$. We already know that the derivative of a constant is equal to zero."

Igor displayed the results of our work:

| Function | Derivative |
|----------|------------|
| $y = c$ | $y' = dy/dx = 0$ |
| $y = cx$ | $y' = dy/dx = c$ |
| $y = f(x) + g(x)$ | $y' = dy/dx = f'(x) + g'(x)$ |
| $y = cx^n$ | $y' = dy/dx = cnx^{n-1}$ |

"With these rules we can find the derivative of any polynomial," Recordis said. "Remember functions like $x^2 + 3x - 5$ or $2x^4 - 3x^3 + 2x^2 - 1$? I think that this just about wraps up the subject of calculus." (Recordis thinks that any problem that can't be expressed using polynomials is not worth bothering with.)

Just as he was saying these words, there was a loud thud out in the courtyard, and the next thing we knew an ominous figure had darted in through the window. Everyone in the room cowered in fear. Although I had never seen the strange apparition before, I could tell from the start that he meant trouble.

"So you think you can outwit me, do you?" he exclaimed, his voice ringing with wicked laughter.

"That's the gremlin," the professor whispered to me. "The Spirit of Hopelessness and Impossibility. He is our arch-enemy."

"You have no idea what you are getting yourselves into," the gremlin cried. "Just wait. First, with your last rule, you have never even thought about what happens if *n* is a fraction. Ah, but even that is too simple. I can show you curves that you have no hope of unraveling."

He held out his cape, and in it we could see misty pictures of strangely oscillating curves of every conceivable shape, which appeared to be floating in space. They seemed to be trying to reach out and strangle us. "What about any of these?" he cried, and a whole chain of algebraic symbols flew out in the air past us.

"This time I am sure to win!" he laughed, as he slowly folded his cape and flew out the window.

## Note to Chapter 2

It is important to note that the derivative can be defined for a particular function only if the limit

$$\lim_{\Delta x \to 0} \frac{f(x + \Delta x) - f(x)}{\Delta x}$$

has a definite value. Some functions, such as $y = |x|$ (the *absolute value*, defined by $y = x$ for $x \geq 0$ and $y = -x$ for $x < 0$), will not have derivatives defined at all points of the function. In this case, the function has no derivative at the point where $x = 0$ (Figure 2–7).

In general, any function with a cusp in it, like the ones in Figure 2–8, will not have a derivative defined at the point where the cusp is located.

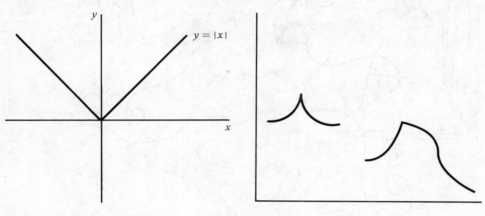

Figure 2–7.                           Figure 2–8.

## Exercises

Find the derivatives of the following functions:

**1.** $y = 3x^3 + 2x^2 + x + 5$

**2.** $y = 4x^5 + x^2$

**3.** $y = ax + 4$

**4.** $y = x^{35}$

**5.** $y = \frac{1}{3}x^3 + \frac{1}{2}x^2 + x + 1$

**6.** $y = (4.34)x^2 + (0.98)x$

**7.** $f(t) = (2t - 5)(3t + 4)$

**8.** Write an expression for $f(x + \Delta x) - f(x)$ for the function $f(x) = ax^2 + bx + c$. Then use the definition of the derivative to find $f'(x)$.

**9.** Using the definition of the derivative, show that, if $y = f(x) + g(x) + h(x)$, then $dy/dx = f'(x) + g'(x) + h'(x)$.

**10.** When Mongol throws his beach ball straight up in the air, its height $h$ at time $t$ is given by $h = -\frac{1}{2}gt^2 + v_0t$. (a) Find the velocity of the ball at time $t$. (b) Find the velocity of the ball when $t = 0$. (c) Find out how long the ball takes to reach its highest point (i.e., at what value of $t$ does $dh/dt = 0$?).

**11.** When Mongol drops his ball off the Hasselbluff Mountain Viewpoint, its height above the ground at time $t$ is given by $h = 64 - \frac{1}{2}gt^2$. (a) What is the velocity at time $t$? (Is the velocity positive or negative?) (b) How long will the ball take to hit the ground (i.e., at what value of $t$ does $h = 0$)? (c) How fast is the ball going the instant before it hits the ground? (d) The quantity $g$ is known as the *acceleration of gravity* and is measured in meters per second². Find a numerical value for $g$ if the ball takes 3.61 seconds to fall to the ground.

12. Find the values of $x$ where the slope of the curve $y = \frac{1}{3}x^3 - x^2 + 3x + 5$ is equal to 3.

13. For what values of $b$ is the line $y = 8x + b$ tangent to the curve $y = x^3$?

14. The *mean value theorem* states that, if a function $y = f(x)$ has a derivative defined everywhere between $x = a$ and $x = b$, then there is some value of $x$ (call it $x_0$) such that $a < x_0 < b$ and $f'(x_0)$ equals the slope of the secant line between the points $(a, f(a))$ and $(b, f(b))$. Consider the function $f(x) = -x^2 + 10x - 15$, and two points on the graph of that function: $(2, 1)$ and $(6, 9)$. Find the value of $x_0$ that is predicted by the mean value theorem (i.e., find $x_0$ such that

$$f'(x_0) = \frac{f(b) - f(a)}{b - a}$$

for $a = 2$ and $b = 6$).

15. The function

$$h(x) = \frac{3x^2 + x - 4}{x - 1}$$

is undefined for $x = 1$. *L'Hôpital's rule* makes it possible to calculate the limit of $h(x)$ in this case. L'Hôpital's rule states that, if $h(x) = f(x)/g(x)$, and $\lim_{x \to a} f(x) = 0$ and $\lim_{x \to a} g(x) = 0$, then $\lim_{x \to a} h(x) = \lim_{x \to a} f'(x)/\lim_{x \to a} g'(x)$. Use L'Hôpital's rule to calculate

$$\lim_{x \to 1} \frac{3x^2 + x - 4}{x - 1}$$

16. *Newton's method* provides an iterative method for estimating the $x$ intercept of complicated functions. The goal of the method is to find $x_0$ such that $f(x_0) = 0$. First, make a guess $(x_1)$ that is reasonably close to the true value of $x_0$. Then calculate a better guess according to the formula $x_2 = x_1 - f(x_1)/f'(x_1)$. The method can be repeated to yield a still better guess, $x_3 = x_2 - f(x_2)/f'(x_2)$. Keep going until you are satisfied that the result is close enough to the true answer. Now, use Newton's method to estimate the cube root of 7. Start with $x_1 = 2$, and find the $x$ intercept of the function $f(x) = x^3 - 7$. Perform a total of three iterations, and compare the result with the true value.

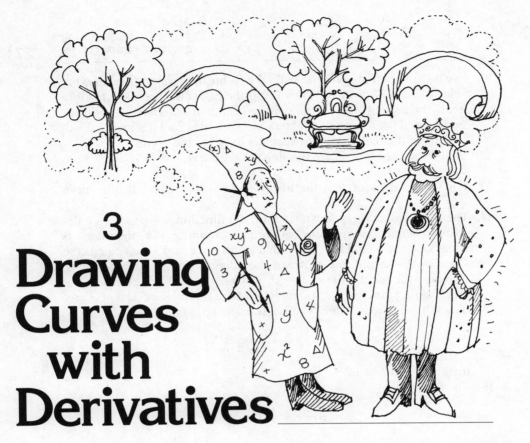

# 3
# Drawing Curves with Derivatives

We were all uneasy as we met in the Main Conference Room the next day. Everyone was still shaken by the visit of the gremlin.

"I have a problem you stuck me with a long time ago," Recordis said. "When we were doing algebra, we decided that we would carve sculptures of various curves in the Royal Garden. You left me with the job of determining the shapes of these curves. If you have ever tried to draw curves involving $x$'s to the third power by plotting points, you will appreciate the problems I have been having."

"How far have you gone?" the king asked.

"The curve you gave me was $y = x^3 + 3x^2 - 9x + 3$. I've figured out four points so far."

"Four points!" the king exclaimed. "Is that all?"

"They don't seem to fit any pattern at all!" Recordis moaned. "I've checked each calculation at least a hundred times to make sure that I haven't made an error, but I can check them again if you want me to."

He told us the four points he had come up with: $(-5, -2)$, $(-2, 25)$, $(0, 3)$, and $(2, 5)$.

"We can make an approximate sketch of the curve and see what it looks like," the professor suggested.

"But we can't even do that!" Recordis said. "We don't know where the curve turns around."

"Turns around?" the king asked.

"It must turn around somewhere," Recordis said. "Sometimes it seems to be going up, sometimes it seems to be going down, and then it seems to be going up again. It must turn around somewhere in the middle."

"Why don't we check about ten points and see what it does?" the king suggested. "That should give us a better idea than just four points."

After some tedious arithmetic, we came up with ten points. (See Table 3–1.)

"See?" Recordis said. "The curve has to turn around."

"There must be some way to figure out where it turns," the professor insisted. (See Figure 3–1.)

"I think I know how we can figure out where it turns," Trigonometeris said. "Let's look at the part of the curve near where it turns." (Figure 3–2.)

**Table 3–1**

| $x$ | $f(x) = x^3 + 3x^2 - 9x + 3$ |
|---|---|
| −5 | −2 |
| −4 | 23 |
| −3 | 30 |
| −2 | 25 |
| −1 | 14 |
| 0 | 3 |
| 1 | −2 |
| 2 | 5 |
| 3 | 30 |
| 4 | 79 |

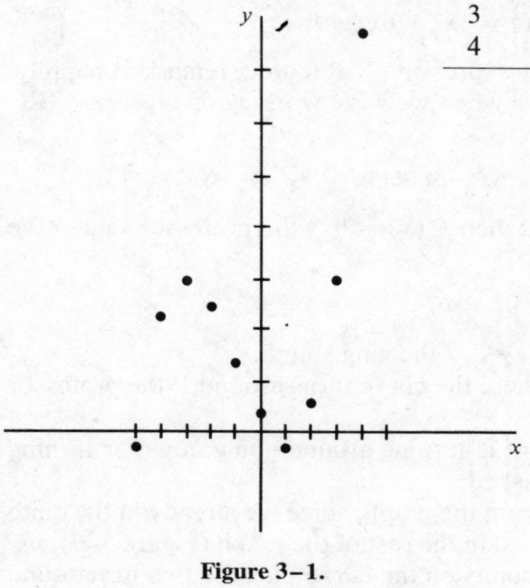

Figure 3–1.

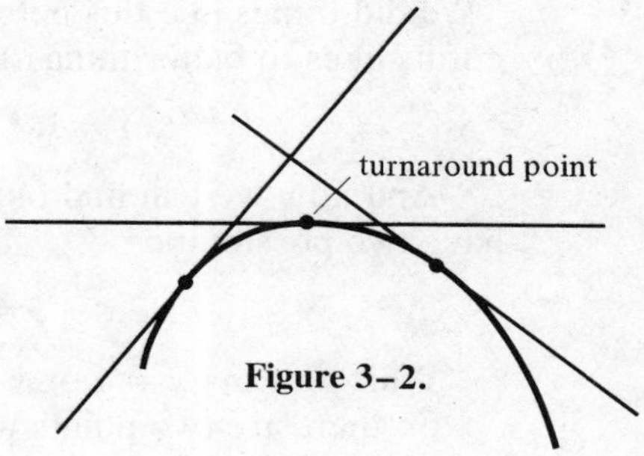

turnaround point

Figure 3–2.

"If the curve is going up," he continued, "then the tangent line to the curve slopes up."

"And its slope is positive," the professor interrupted.

"If you keep drawing lines closer and closer to the turnaround point, they gradually become flatter. At the exact point where the curve turns around, the tangent line must be parallel to the horizontal $x$ axis."

"Of course!" the professor said. "And it must have a slope of zero, just like any other horizontal line."

"And we know what the slope of the tangent line at a given point is," Trigonometeris contributed.

"Of course we know that!" the professor said. "The value of the derivative at a given point is the slope of the tangent line at that point. To find the places where the curve turns around, all we need to do is find the places where the derivative is zero."

"Of course!" Recordis exclaimed. "Why didn't I think of that?"

"And the same thing happens when the curve is going down, and then starts turning around to go up," Trigonometeris said.

"Let's find the derivative of Recordis' curve," the king suggested. "We can use what we did yesterday."

$$f(x) = x^3 + 3x^2 - 9x + 3$$

"The derivative of the first term is $3x^2$," the professor said.

"The derivative of the second term is $6x$," the king said.

"The derivative of the third term is $-9$," Trigonometeris said.

"And the derivative of the last term is zero," Recordis said. "That gives us the derivative of the whole polynomial."

$$f'(x) = 3x^2 + 6x - 9$$

"I think we can factor that expression," Recordis remarked happily. "We did things like this before when we were working on algebra." (Recordis likes to factor things.)

$$f'(x) = 3x^2 + 6x - 9 = (3x - 3)(x + 3)$$

"And now we can find out when $f'(x) = 0$," the professor said. "We have two possibilities."

$$3x - 3 = 0 \quad \text{or} \quad x + 3 = 0$$

"Which means $x = 1$ or $x = -3$," the king stated.

"So there are two points where the curve turns around," the professor said.

"How can we tell whether it is turning around going down or turning around going up?" Recordis asked.

"We can tell pretty easily from the graph, since we already know quite a few points." Igor quickly filled in the rest of the graph (Figure 3–3), using what we knew about the points on the curve plus the two turnaround points.

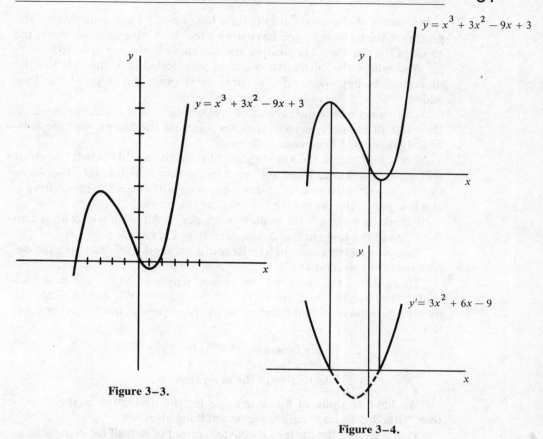

**Figure 3–3.**

**Figure 3–4.**

"It is obvious that the curve reaches a high point when $x = -3$, and then starts going down again," the professor said. "When $x = +1$, the curve reaches a low point and starts going up again."

"But look at how many points we had to calculate before we had any idea about the shape of the curve," Recordis protested. "You always make me do all the work. We had better figure out a quick way to tell whether a point with a horizontal tangent is a high point or a low point. I simply refuse to calculate ten points each time."

"We should be able to figure out something," the professor said.

After we had looked at the graph for a long time, I came up with a suggestion. "Let's look at the graph of the curve and its derivative together."

"How can we do that?" Recordis asked.

"When we take the derivative of some function, we end up with another function. In this case, we have two functions: $f(x) = x^3 + 3x^2 - 9x + 3$ and $f'(x) = 3x^2 + 6x - 9$. There is no reason why we can't draw a graph of $f'(x)$ just the same as we drew a graph of $f(x)$. Why don't we draw the graph of the derivative right under the graph of the main function?" (Figure 3–4.)

We stared at these two graphs for a long time. "I see something," the professor said. "Where you have drawn the derivative as a solid line, the value of the derivative is positive and the curve is sloping upward."

"And where the derivative is drawn as a dotted line, the value of the derivative is negative and the curve is sloping downward," the king added.

"The places where the derivative crosses the $x$ axis are the ones where the value of the derivative is zero, so these are the places with the horizontal tangents," Trigonometeris said.

After a long pause the king began, "I want to say this slowly, or else I will make a mistake. When the main curve is at a high point, it looks as though the derivative curve is sloping downward. When the main curve is at a low point, the derivative curve is sloping upward."

"I think he's right," the professor gasped. "All we have to do is find the slope of the tangent line to the derivative curve."

There was a brief pause before Recordis shouted, "We can take the derivative of the derivative!"

"There doesn't seem to be any reason why we can't," the professor said. "We have the function $f'(x) = 3x^2 + 6x - 9$. We should be able to find the derivative of that function as easily as we can for any other function."

> main function: $x^3 + 3x^2 - 9x + 3$
> derivative: $3x^2 + 6x - 9$
> derivative of the derivative: $6x + 6$

"We have to think of a better name for the derivative of the derivative," Recordis said, "before we do anything else."

"That's easy," the professor boasted, proud of herself for being able to think of a name without asking me. "We'll call it the *second derivative*."

"That means we can call the derivative of the main function the *first derivative*, if we want to," the king said.

"And if we wanted to we could also have the third or fourth or fifth derivative," Trigonometeris added.

"What would we ever do with so many derivatives?" Recordis asked. "I think we have too many derivatives as it stands now. We do need to think of a symbol for the second derivative, though."

"That should be easy," the professor said. "If the first derivative has one prime after it, the second derivative should have two primes after it."

> function: $y = f(x)$
> first derivative: $y' = f'(x)$
> second derivative: $y'' = f''(x)$

"I like your $dy/dx$ system," the king stated. "We should have a way of naming second derivatives that uses that notation."

"I have an idea," I said. "Maybe we can clarify what the expression

$dy/dx$ means. We start with a function $y$. Then we hit the $y$ with the expression $d/dx$, which gives us the derivative: $dy/dx$. It looks as though the symbol $d/dx$ stands for the *operation* of taking the derivative with respect to $x$.''

"That sounds reasonable," Recordis agreed. "I'd rather write $d/dx$ than 'take the derivative of the quantity with respect to $x$.' "

"So let's see what happens if we take the operator $d/dx$ and apply it to the first derivative: $dy/dx$," I said.

"You get $(d/dx)(dy/dx)$," the professor told us.

"It looks as though you're multiplying two fractions," the king said.

"It looks like that, but that's not what you're doing," the professor stated.

"I think we can make that notation even shorter," Recordis said. "Let's write $d\ dy/dx\ dx$."

"We can make it shorter still," Trigonometeris added. "Instead of writing that $d$ on the top twice, we may just as well write $d$ with a 2 after it, like this: $d^2y/dx\ dx$."

"We can also save ourselves the trouble of writing $dx$ twice," the professor said. "We can write $d^2y/dx^2$."

Recordis was hoping we could shorten that even further, but we all agreed that it was the best we could do. Igor displayed our new notation:

function: $y = f(x)$
derivative: $y' = f'(x) = dy/dx$
second derivative: $y'' = f''(x) = d^2y/dx^2$

"Now we can have Igor draw the graphs of all three of these functions," the professor said. (Figure 3–5.)

"We can make the following rule," she continued. "If the first derivative is zero and the second derivative is *positive*, the main function reaches a *low* point. If the first derivative is zero and the second derivative is *negative*, the curve reaches a *high* point."

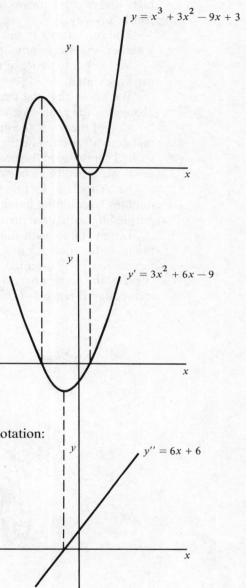

$y = x^3 + 3x^2 - 9x + 3$

$y' = 3x^2 + 6x - 9$

$y'' = 6x + 6$

**Figure 3–5.**

"That should be easy to remember," Trigonometeris said. "You just remember that it is backwards from the way it should be. Negative means high, and positive means low. I always knew algebra was somewhat backwards anyway."

"The rule works in this case," Recordis said, "but how do you know whether it works in any other case?"

"We've got to figure out what the second derivative means in general," the king stated.

"We know that the derivative is the rate at which the main function changes," the professor said. "For example, if the function represents the position of the train, the first derivative represents its velocity. So the second derivative must be the rate at which the first derivative changes." She looked shrewdly at the board. "I have an idea. Igor, give me a bug with an arrow strapped to its back."

The Visiomatic Picture Chalkboard Machine looked startled, but a few minutes later a little hatch opened in the side of the screen and out flew a purple bug with a red arrow strapped to its back.

"I want you to walk along this function," the professor said, pointing to the main function $f(x) = x^3 + 3x^2 - 9x + 3$, which was still sketched on the screen. She oriented the bug so that the arrow was tangent to the curve at the place where the bug was standing, and pointing in the positive $x$ direction (Figure 3–6).

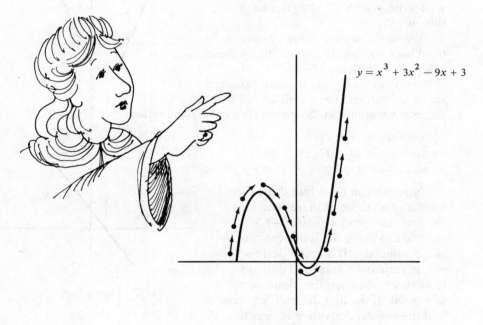

$$y = x^3 + 3x^2 - 9x + 3$$

**Figure 3–6.**

"Amazing!" Trigonometeris gasped. "I have never figured out how bugs are able to walk on walls."

"Look at what's happening to the arrow," the professor said as the bug walked along the function.

"It's slowly getting flatter," the king observed.

"Of course the arrow is getting flatter when the curve is turning downward," Recordis said. "What is that supposed to prove?"

"You just said it!" the professor exclaimed, a sudden light coming into her eyes. "When the curve is turning downward, the slope of the tangent line is becoming less."

"Right."

"When the value of the first derivative is becoming less, the first derivative curve must be sloping downward."

"Right."

"And when the first derivative curve is sloping downward, its slope must be negative."

"Right."

"And when the slope of the first derivative curve is negative, the value of the second derivative must be negative," the professor finished triumphantly.

"Can you go through that again?" the king asked.

"What we just said is that, when the value of the second derivative is *negative,* the main curve is *turning downward,*" the professor told him.

"That means it must look like one of these curves," Recordis said. Igor drew a lot of curves turning downward (Figure 3–7).

"We invented a name for a shape like that," the king reminded us. "Remember the time Mongol was running along that hill and it caved in? It made a giant hole, so we made up the name *concave* for something shaped like a hole."

"We could say that these curves (Figure 3–7) are oriented so that their concave part is downward," the professor said.

"I know how you can draw curves that are concave upward," the king stated. Igor obligingly drew some curves that were concave upward (Figure 3–8).

"When a curve is concave upward, you can see exactly the opposite happen," the professor said. The little bug continued walking along the

**Figure 3–7.**

**Figure 3–8.**

original function until it reached the point where the curve was concave upward. "As you move along the curve, the tangent arrow slowly turns upward. That means that the slope of the tangent line is increasing, so the value of the second derivative must be positive. We can summarize these rules: When the second derivative is *positive,* the main curve is *concave upward.* When the second derivative is *negative,* the main curve is *concave downward.*"

"I'm still confused trying to remember which way is concave and which way is not concave," Recordis confessed.

"I know what we can say," the king said. "If you have a curve that is concave upward, you could easily pour water into it. If you have a concave-downward curve, the water would spill right out of the curve." (Figure 3–9.)

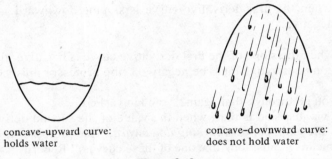

concave-upward curve:
holds water

concave-downward curve:
does not hold water

**Figure 3–9.**

"I think I can remember that," Recordis said. Igor made a table:

second derivative is positive—curve holds water (concave upward)
second derivative is negative—curve spills water (concave downward)

"That's back the way it should be," Trigonometeris said. "Positive means up, and good (if you're trying to save water). Negative means down, and bad."

"Now we can tell whether a point is a minimum or a maximum," the professor continued. "If the curve is concave downward, then the point with a horizontal tangent is at the top of the curve, so to speak." (Figure 3–10.)

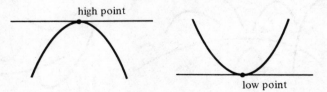

high point

low point

**Figure 3–10.**

"The opposite thing happens with a concave-upward curve," the king said. "The point with a horizontal tangent is then a low point."

Igor summarized:

$y' = 0$, $y''$ positive, point is a minimum
$y' = 0$, $y''$ negative, point is a maximum

"Wait a minute," the professor said. "Look at this curve (Figure 3–11). You can't say that point $A$ is a maximum. The value of the curve at point $B$ is greater than the value of the curve at point $A$."

"I think you could call point $A$ a *local maximum*," I suggested. "Farmer Floran told me that Crabgrass Hill is the highest point in Coast City, but that the mountains off in the country are higher than the hill. There's no reason why you can't call a point a local maximum if it is higher than all the points around it."

"I suppose you could call the highest point a curve ever reached the *absolute maximum*," the professor suggested.

"Our original curve doesn't have an absolute maximum," Recordis pointed out. "It goes off to infinity."

"I can think of a curve that does have an absolute maximum," the king said. Igor drew the graph of $y = -x^2$ (Figure 3–12).

"The point $(0, 0)$ is the highest point this curve ever reaches, no matter how far you extend the curve in either direction," the king went on. "That means that $(0, 0)$ is an absolute maximum."

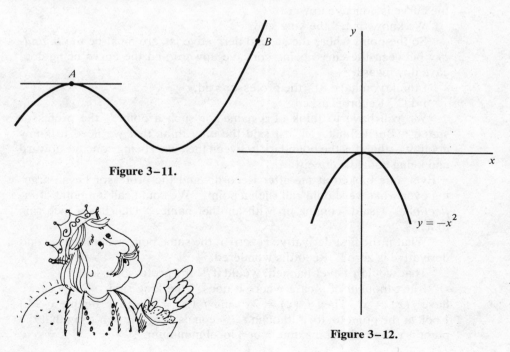

Figure 3–11.

Figure 3–12.

"Let's make sure that this curve matches our theories," the professor cautioned. "If $y = -x^2$, then $y' = -2x$, and $y'' = -2$. The second derivative is negative all the time, so the curve should be concave downward all the time." The professor looked carefully at the graph.

"Of course it is concave downward," Recordis said. "You can see that it would spill water."

"I just thought of something else," the king stated. "What happens if the second derivative is zero?"

"Fiddlesticks!" the professor said. "I knew somebody would think of a complication."

We looked at the first graph again.

"The value of the second derivative is zero when $x = -1$," the professor said.

"On the original curve, that happens at the point $(-1, f(-1))$," the king added.

"What does that mean?" the professor asked.

There was a long pause before Recordis began slowly. "I'll say what it looks like to me, but I don't think this will help much."

"Go ahead," the professor said.

"When $x$ is less than $-1$, the second derivative is negative. That means that the curve is concave downward when $x$ is less than $-1$."

"We know that," the professor said.

"And when $x$ is greater than $-1$, the second derivative is positive, so the curve is concave upward."

"We know that," the king said.

"So the point where the second derivative is zero must be the *boundary* between the curve being concave upward and the curve being concave downward."

"I think you have it!" the professor said.

"I do?" Recordis asked.

"We will have to think of a name for such a point," the professor stated. "But I think you just said the only thing that we need to know about it—that it is the boundary between the curve being concave upward and being concave downward."

Everyone looked at me after Recordis and the professor began arguing over what we should call such a point. "We could call it a point of *inflection*," I said, coming up with another name. "That has a nice ring to it."

"What if the first derivative is zero at the same point where the second derivative is zero?" Recordis wondered.

"That wouldn't ever happen, would it?" the professor asked.

"I just thought of a case where it does," the king said. "Suppose we have $f(x) = x^3$. Then $f'(x) = 3x^2$ and $f''(x) = 6x$. (See Figure 3–13.) Look at the point $(0, 0)$. Although $f'(0)$ equals zero, we can't say that the point is either a local maximum or a local minimum."

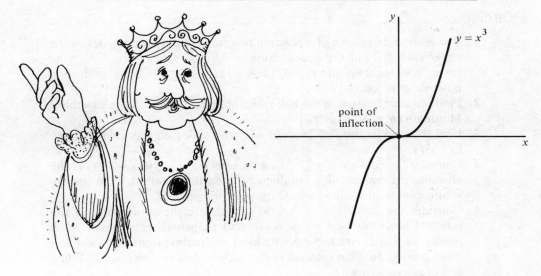

**Figure 3–13.**

"The point (0, 0) is still a point of inflection, though," the professor said. "When $x$ is less than zero the curve is concave downward, and when $x$ is greater than zero the curve is concave upward."

"It still is a point with a horizontal tangent," Trigonometeris noted.

"We can summarize all the rules we've come up with for sketching curves," the professor said. "This subject might be useful for more than just finding the speed of the train."

---

## RULES FOR CURVE DRAWING

1. When the first derivative is positive, the value of the function is increasing.
2. When the first derivative is negative, the value of the function is decreasing.
3. When the first derivative is zero, the curve has a horizontal tangent at that point.
4. When the second derivative is positive, the curve is concave upward (and it holds water).
5. When the second derivative is negative, the curve is concave downward (and it spills water).
6. When the second derivative is zero, the curve has a point of inflection.
7. When the first derivative is zero and:
   (a) the second derivative is positive, the point is a local minimum.
   (b) the second derivative is negative, the point is a local maximum.
   (c) the second derivative is zero, the point is a point of inflection with a horizontal tangent.

## Exercises

1. The second derivative of a position function is known in physics as the *acceleration*. Find the acceleration of Mongol's beach ball when he tosses it straight up into the air ($h = -\frac{1}{2}gt^2 + v_0t$). Is the acceleration positive or negative?

2. Find the acceleration of the ball when Mongol throws it off Hasselbluff Mountain ($h = 64 - \frac{1}{2}gt^2$).

3. Find the acceleration of Mongol when he slides on Ice Skating Lake ($x = 3t$).

4. Consider the curve $y = x^2 - 3x$ on the interval $x = 0$ to $x = 5$. Find the absolute maximum value obtained by $y$ on this interval. Find the absolute minimum value obtained by $y$ on this interval.

5. Consider the curve $y = -x^4 + 8x^2$. What are the coordinates of the points where the curve has horizontal tangents? How many such points are there? Are these points local maximum points or local minimum points? In what interval is the curve concave downward? When is it concave upward?

6. Consider the curve $y = x^2 - \frac{1}{3}x^3$. Where is the curve rising? Where is it falling? Where is it concave upward? Where is it concave downward?

7. $y = x^3 + x^2 + x$; $d^4y/dx^4 = ?$

8. $y = ax^3 + bx^2 + cx$; $d^3y/dx^3 = ?$

9. $y = x^n$; $d^ny/dx^n = ?$

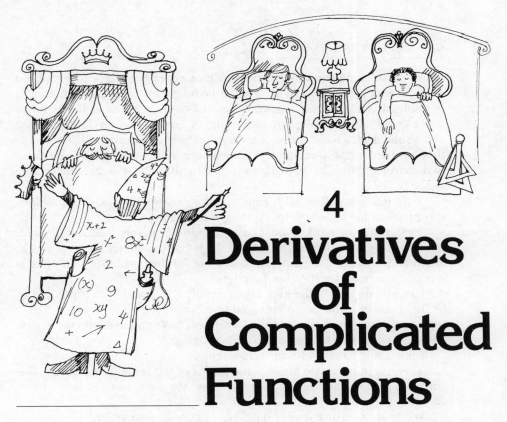

# 4
# Derivatives
# of
# Complicated
# Functions

Recordis woke everyone early the next morning. "I remember another problem!" he shouted. "I think we can solve it easily."

Everyone was still sleepy as we gathered in the Main Conference Room. Recordis had Igor draw a map. (See Figure 4–1.)

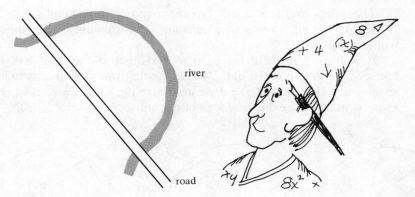

river

road

**Figure 4–1.**

"You remember the field I have out in the country," he began. "There is a stream that runs in a perfect half-circle around one end, and there is a straight road that borders it on the other end. I've been wanting to build a house on the field for a long time, but I've never been able to figure out what its dimensions should be. I want the house to cover as much area as possible. It has to have square edges, of course, because the builders don't know how to make any other kind of house. It is obvious that you can't build a very wide house or a very thin house (Figure 4–2). The shape that has the largest area must be somewhere in the middle between these two extremes. We have to figure out exactly what the maximum-area shape is.''

''Why did you call us all together?'' the professor asked. ''What does this problem have to do with anything we have been doing?''

''Yesterday we worked on finding maximum points,'' Recordis said. ''All we have to do is find the area of the house as a function of its width, and then we can find the derivative to solve for the particular width that is a maximum point for the area function.''

''Can you do that?'' the king asked timidly. ''I thought functions had to be functions of $x$ that were set equal to $y$.''

''I don't see why that has to be,'' I replied. ''We can make up some new letters to stand for the variables in the problem.''

''Like what?'' the professor said sleepily.

''The first thing we do know is the radius of the semicircle, right?'' Recordis didn't say anything. ''We do know the radius of the semicircle, don't we?''

''We could measure it very quickly,'' Recordis suggested.

''We don't need to know the exact length while we're doing the calculation,'' I said. ''For now we will say that the length of the radius is equal to $r$.''

''It should be easy to remember,'' Recordis said, ''that $r$ stands for radius.''

''The radius won't change,'' Trigonometeris offered helpfully.

''Right,'' I said, ''$r$ will be a constant. Unfortunately, everything else will be a variable.''

''It would be easier if we tried to maximize the area of half of the house,'' the king pointed out. ''We know that the total area is twice the shaded area (Figure 4–3), so if we maximize the shaded area it will be the same as maximizing the total area of the house.''

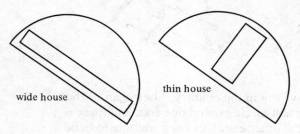

Figure 4–2.

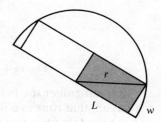

Figure 4–3.

"The shaded area will be where the living room will be," Recordis said.

"We'll call the length of the living room $L$ and the width of the living room $w$," I stated. "Then we know that (area of living room) $= w \times L$."

"We can find a relationship between $w$ and $L$," the professor said. "$L$ and $w$ will always form the sides of a right triangle, with the radius of the circle forming the hypotenuse. Then we can use the Pythagorean theorem to show that $L^2 + w^2 = r^2$."

"Now we can solve for $L$," Recordis said.

$$L^2 + w^2 = r^2$$

$$L^2 = r^2 - w^2$$

$$L = \sqrt{r^2 - w^2}$$

"And we can put the last expression in the equation for $A = wL$," the king added.

$$A = w\sqrt{r^2 - w^2}$$

"That's great!" Recordis exclaimed. "We now have $A$ expressed as a function of only one variable: $w$. That means $A = f(w)$. All we need to do now is find $dA/dw$, and then set $dA/dw = 0$. Then we can solve for the value of $w$ that maximizes $A$."

"Hold it!" the professor objected. "We can't find the derivative of that function! We don't know whether the $f(x) = x^n$, $f'(x) = nx^{n-1}$ rule works if $n$ is a fraction, and in this case we have something raised to the $\frac{1}{2}$ power. And we don't even have just $w$ raised to a power; we have a weird expression like $(r^2 - w^2)$ raised to a power. Not only that, but we have a stray $w$ in front of the complicated function that is multiplying everything else! There is no way we can do that problem!"

"You're right!" Recordis said, looking closely at the function. "That is hopeless. That is really impossible."

"Then why did you get us up so early?" the professor complained. Everyone began to get ready to go back to bed.

"There has to be something we can do!" the king said.

Everyone stopped. "It looks to me as though we have three problems," I said. "First, we have that loose $w$ in front of the rest of the function. We know that it is pretty easy to find the derivative of a function if we have a constant number multiplied by a function. We just need to find a new rule that tells us what to do if we have a variable times a function. Then we need to see what to do about an embedded function; for example, in this case instead of having $w^{1/2}$ we have $(r^2 - w^2)^{1/2}$. All this means is that we have one function of $w$ $(f(w) = r^2 - w^2)$ embedded in another function $(g(q) = q^{1/2})$, giving us $g(f(w))$. We need to figure out what to do with a function like that. Finally, we need to see whether the rule for finding the derivative of a power works when we have fractional powers like square roots. If we can just solve these three problems, we can find the answer to the problem we started with."

"Right," the professor said. "And if I could just grow 20 feet and develop some huge muscles, I could beat Mongol in a wrestling match."

"We might as well at least give it a try," the king stated.

"The first problem is what to do when we have two functions multiplied together. Suppose we have two functions, $u(x)$ and $v(x)$, and a function $q(x)$ that is set up so that $q(x) = u \times v$. We need to find out what $dq/dx$ is."

"I know what would be really simple," Recordis said. "When we had $q(x) = f(x) + g(x)$, we just said that $dq/dx = df/dx + dg/dx$. Why don't we say that, if $q(x) = uv$, then we have $dq/dx = (du/dx)(dv/dx)$?"

"That would be too easy," the king objected.

"Besides, we can't just make something up like that," the professor said.

"I thought we were making most of this up anyway," Recordis protested.

"I don't think what you suggested will work," I said. "Suppose we had the function $q = 10x$. We know that

$$\frac{d}{dx}(x) = 1 \quad \text{and} \quad \frac{d}{dx}(10) = 0$$

If

$$\frac{dq}{dx} = \frac{d(10)}{dx}\frac{dx}{dx}$$

we would get $dq/dx = (1)(0)$, which is equal to zero."

"That won't work because we know that in this case $dq/dx = 10$," the professor said.

"All right, we can look for a new rule," Recordis agreed. "I was just trying to make life simple for all of us."

I suggested, "First, let's see whether we can find out what $q(x + \Delta x)$ is."

"That would be this expression," the professor said.

$$q(x + \Delta x) = u(x + \Delta x)\, v(x + \Delta x)$$

"We can rename $u(x + \Delta x)$," I said. "Show us a graph of some function, Igor." (Figure 4–4.)

"We could call the difference between $u(x)$ and $u(x + \Delta x)$ something like $\Delta u$," the king suggested.

$$u(x + \Delta x) - u(x) = \Delta u$$

"We can call the difference between $v(x + \Delta x)$ and $v(x)$ something like $\Delta v$," the professor said. "Then we can write two equations."

$$u(x + \Delta x) = u(x) + \Delta u$$

$$v(x + \Delta x) = v(x) + \Delta v$$

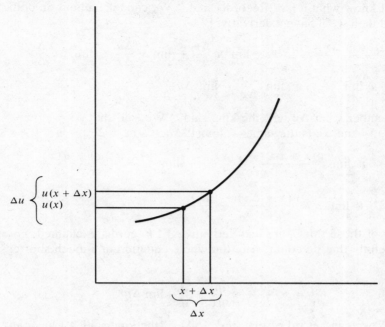

**Figure 4–4.**

"We can put these expressions back in our equation for $q(x + \Delta x)$," the king noticed.

$$q(x + \Delta x) = u(x + \Delta x)\, v(x + \Delta x)$$

$$= (u(x) + \Delta u)(v(x) + \Delta v)$$

"We can multiply out the right side algebraically," Recordis offered.

$$q(x + \Delta x) = u(x)\, v(x) + u\, \Delta v + v\, \Delta u + \Delta u\, \Delta v$$

"I know what $u(x)\, v(x)$ is!" the king said. "That's equal to $q(x)$."

$$q(x + \Delta x) = q(x) + u\, \Delta v + v\, \Delta u + \Delta u\, \Delta v$$

"We can subtract $q(x)$ from both sides," the professor said. She was beginning to get the gleam in her eye that she gets when she thinks she sees something that Recordis hasn't noticed yet.

$$q(x + \Delta x) - q(x) = u\, \Delta v + v\, \Delta u + \Delta u\, \Delta v$$

"That's beginning to look familiar," Recordis stated. "I just can't think of what it looks like."

"Let's divide both sides by $\Delta x$," the professor suggested.

$$\frac{q(x + \Delta x) - q(x)}{\Delta x} = u\, \frac{\Delta v}{\Delta x} + v\, \frac{\Delta u}{\Delta x} + \Delta u\, \frac{\Delta v}{\Delta x}$$

"Now I know what it is!" Recordis said. "We can take a limit on both sides, and then we'll have a derivative!"

$$\lim_{\Delta x \to 0} \frac{q(x + \Delta x) - q(x)}{\Delta x} = \lim_{\Delta x \to 0} u \frac{\Delta v}{\Delta x} + \lim_{\Delta x \to 0} v \frac{\Delta u}{\Delta x} + \lim_{\Delta x \to 0} \Delta u \frac{\Delta v}{\Delta x}$$

$$\frac{dq}{dx} = u \lim_{\Delta x \to 0} \frac{\Delta v}{\Delta x} + v \lim_{\Delta x \to 0} \frac{\Delta u}{\Delta x} + \lim_{\Delta x \to 0} \Delta u \frac{\Delta v}{\Delta x}$$

"I remember what $\Delta v$ is," the king said. "We said that $\Delta v = v(x + \Delta x) - v(x)$. And $\Delta u$ is the same—almost; $\Delta u = u(x + \Delta x) - u(x)$."

$$\frac{dq}{dx} = u \lim_{\Delta x \to 0} \frac{v(x + \Delta x) - v(x)}{\Delta x} + v \lim_{\Delta x \to 0} \frac{u(x + \Delta x) - u(x)}{\Delta x}$$

$$+ \lim_{\Delta x \to 0} \Delta u \frac{v(x + \Delta x) - v(x)}{\Delta x}$$

"Three of those things are just derivatives!" Recordis exclaimed, noticing gleefully that he could write the whole equation in a much shorter fashion.

$$\frac{dq}{dx} = u \frac{dv}{dx} + v \frac{du}{dx} + \frac{dv}{dx} \lim_{\Delta x \to 0} \Delta u$$

"I think we can get rid of that $\lim_{\Delta x \to 0} \Delta u$," the king said. "I have an idea. Remember that $\Delta u = u(x + \Delta x) - u(x)$. When we take the limit of $\Delta x$ going to zero, we get $\Delta u = u(x) - u(x)$."

"That's equal to zero," the professor pointed out.

"That means we can erase it," Recordis said jubilantly. "If it is equal to zero, it doesn't make any difference whether it is multiplied by $dv/dx$ or 752 or $3,569,204\frac{1}{2}$."

Igor displayed the final rule:

$$q(x) = uv$$

$$\frac{dq}{dx} = u \frac{dv}{dx} + v \frac{du}{dx}$$

"That isn't nearly as bad as I thought it would be," Trigonometeris said.

"We still had better think of a name for it," Recordis told him. "It will be hard enough to remember as it is."

"The name will be easy," the professor said. "We can use this rule to find the derivative of a function if it is the product of two functions, so we'll call it the *product rule*. And I can think of a sentence that will make it a little bit easier to remember. To find the derivative of two functions multiplied together, take the first function, multiply it by the derivative of the second function, and add the result to the second function multiplied by the derivative of the first function."

"Let's make sure that this rule agrees with what we did before," Recordis suggested. "It would be a tremendous embarrassment if we derived a new rule like this and it turned out to give us a different result for some problem to which we already know the answer."

"We already found the result in a simple case where two functions are multiplied together," the king said. "We had $y = cx$, where $c$ was a constant. Let's try the product rule on that function."

"We can try," the professor said nervously. "First, we need the first function. That's $c$. Then we need the derivative of the second function. That's 1. That gives us 1 times $c$ as the first term. Then we need to add the second function [the sweat was beginning to build on her brow], which is $x$, multiplied by the derivative of the first function, which is the derivative of $c$, which is . . ."

"Zero!" the king supplied. "We end up with this expression."

$$\frac{d(cx)}{dx} = (1)(c) + (0)(x) = c$$

"That is the same answer we got before," the professor said happily. "Our rule does work."

"You were lucky that time," Recordis objected. "I have another function. How about $f(x) = x^2$? We already know that $f'(x) = 2x$. In this case we know that $f(x) = x$ times $x$. Let's try your product rule on that function."

"All right," the professor said, slightly more confident. "We have the first function (which is $x$) times the derivative of the second function (which is the derivative of $x$, which is 1). Then we have the second function (which is also $x$) times the derivative of the first function, which is also 1. That means we end up as follows."

$$f(x) = x \cdot x$$

$$f'(x) = (1)(x) + (x)(1) = 2x$$

"That does equal $2x$!" the king said. "It is consistent with what we did before."

Igor added the product rule to our list.

"One down and two to go," the professor said to me. "I think the other two problems are harder, though."

---

**PRODUCT RULE**

$$f(x) = u(x)\, v(x)$$

$$\frac{df}{dx} = u\,\frac{dv}{dx} + v\,\frac{du}{dx}$$

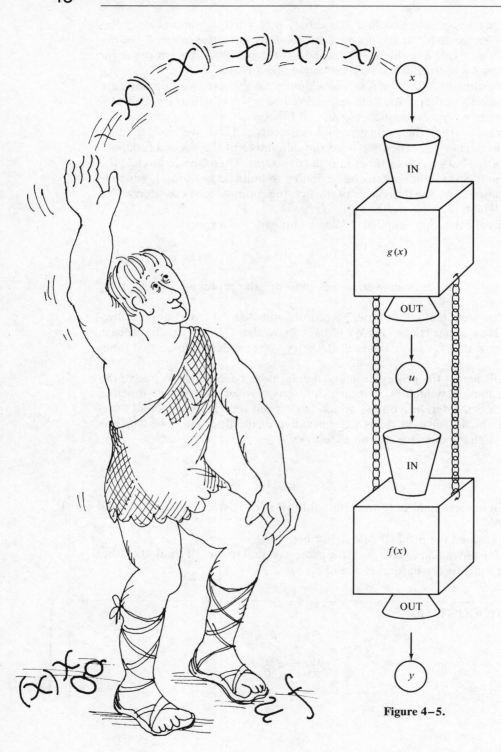

**Figure 4–5.**

"All right," I said. "Now we have the problem of a function embedded in another function. (The mathematical term for a function of this form is *composite function*.) The problem is this: Find $dy/dx$ when $y = f(g(x))$."

"What exactly does $f(g(x))$ mean?" Recordis asked.

"Both $f$ and $g$ are function machines," I said. "All the embedded function means is that we have Mongol throw an $x$ into the $g$ machine. The $g$ machine spits out some number $g(x)$, or we could call it $u$. Then we set up a machine so that the $g(x)$ number $u$ gets thrown right into the $f$ machine, which spits out $y$, which is the number Mongol wanted to get out of the machine in the first place." (Figure 4–5.)

"It looks as though you would need a chain to hold the $f$ machine and the $g$ machine together," Recordis noted.

"Now suppose that Mongol throws an $(x + \Delta x)$ into the machine," I said.

"Then the $g$ machine will spit out the number $g(x + \Delta x)$," the professor offered.

"Let's call that number $(u + \Delta u)$," I said.

$$u + \Delta u = g(x + \Delta x)$$

"Then the number $(u + \Delta u)$ will fall into the $f$ machine," the king said.

"That machine will then throw out the number $f(u + \Delta u)$," I added.

"I bet a hundred dollars you're going to call that number $y + \Delta y$," Recordis said.

$$y + \Delta y = f(u + \Delta u)$$

"We're looking for $dy/dx$," the professor said. "That means that we have to find the limit of $\Delta y/\Delta x$ as $\Delta x \to 0$. I don't see how you're going to do that."

"We can say the following," I suggested.

$$\frac{\Delta y}{\Delta x} = \frac{\Delta y}{\Delta u}\frac{\Delta u}{\Delta x}$$

"You can't just say something like that!" Recordis protested.

"Sure we can," I said. "The little numbers with the deltas are all real numbers, and they can be very, very small but we'll never let them be equal to zero. Then the equation we wrote is valid no matter how we define $\Delta y$, $\Delta x$, or $\Delta u$."

"We could try taking the limits of both sides now," the professor suggested.

$$\lim_{\Delta x \to 0}\frac{\Delta y}{\Delta x} = \lim_{\Delta x \to 0}\frac{\Delta y}{\Delta u}\frac{\Delta u}{\Delta x}$$

"The left-hand side is equal to $dy/dx$," Recordis noted.

$$\frac{dy}{dx} = \lim_{\Delta x \to 0}\frac{\Delta y}{\Delta u}\lim_{\Delta x \to 0}\frac{\Delta u}{\Delta x}$$

"The last part is equal to *du/dx*," the king said.

$$\frac{dy}{dx} = \lim_{\Delta x \to 0} \frac{\Delta y}{\Delta u} \frac{du}{dx}$$

"We found out a few minutes ago that, when $\Delta x$ goes to zero, $\Delta u$ goes to zero," Recordis reminded us.

"We could rewrite our expression like this," the professor said.

$$\frac{dy}{dx} = \lim_{\Delta u \to 0} \frac{\Delta y}{\Delta u} \frac{du}{dx}$$

"I know what that is equal to!" shouted the king.

$$\frac{dy}{dx} = \frac{dy}{du} \frac{du}{dx}$$

"Is that legal?" Recordis asked. "I thought we had to have a *dx* on the bottom all the time."

"I don't see why it has to be that way all the time," the professor answered. "In this case we have $y = f(u)$, so we should be able to say that the derivative is *dy/du*, just the same as we can say that if $y = f(x)$ the derivative is *dy/dx*."

"We must have a good name for this one," Recordis said.

"I know what we can call it," the king said. "In the original design for the composite function machine, we said that we needed a chain to hold the two functions together. We may as well call this rule the *chain rule*."

---

### CHAIN RULE

$$y = f(g(x)) \qquad \text{Let } u = g(x), \text{ so } y = f(u).$$

$$\frac{dy}{dx} = \frac{dy}{du} \frac{du}{dx}$$

---

"I know a good way to remember the rule," the professor offered. "It looks as though derivatives behave as if they are two fractions being multiplied together, and the *du*'s just cancel each other out."

"That's two problems out of three," the king said. "Now we come to the hard one. What do we do if we have a fractional power?"

"Let's start with $y = x^n$, only this time *n* isn't an integer," the professor suggested.

"That doesn't help much," the king said.

"How else can we write it?" the professor asked.

"If it is a fraction, we can write it as a ratio of two unknown integers," I suggested. "We could say that $y = x^{p/q}$."

"Will that work?" the professor demanded. "How do you know that *p* and *q* are integers?"

" 'Any rational number can be expressed as a ratio of two integers,' " Recordis proudly quoted from a page in his book. "That's the definition of a rational number."

"We're still stuck," the professor said. "Remember that an integer in the denominator of an exponent means to take that root of the number. For example, $x^{1/2}$ means the square root of $x$, and $x^{1/3}$ means the cube root of $x$. Now suppose we use the definition of the derivative. Then we get the following:

$$y' = \lim_{\Delta x \to 0} \frac{(x + \Delta x)^{p/q} - x^{p/q}}{\Delta x}$$

"The problem is going to arise when we have the quantity with the addition sign $(x + \Delta x)$ raised to a fractional power. That's the same thing as $\sqrt[q]{(x + \Delta x)^p}$. We don't have any way of evaluating a radical sign if we are adding together two numbers under it! We can't do anything with that kind of expression even if we know what the fractional power is, for example, $\sqrt{a + b}$ or $\sqrt[3]{a + b}$. We can't do anything with either of those expressions! It will be even harder when we don't know what root we're taking—in this case we know only that we're taking the $q$th root. We absolutely cannot do it this way."

"Then we must think of another way," the king stated simply.

$$y = x^{p/q}$$

"I have an idea," I said. "We know that our formula for the derivative of a power works if we have integer powers. That means we must set up the equation so that we are dealing only with integer powers."

"Just try," the professor told me.

"Suppose we take both sides of the equation and raise them to the power $q$," the king suggested.

$$y^q = (x^{p/q})^q$$
$$= x^p$$

"Sure, that gets rid of the fraction," the professor said. "But look at how many more problems it causes! Now we no longer have a function defined explicitly!"

"Do we have to?" I asked.

"Don't we?" Recordis said. "Everything we have done before has had $y = f(x)$. We've never tried to apply a function to the $y$ before."

"But we've run across this kind of thing in algebra," the king noted. "We derived an equation for a circle that was $x^2 + y^2 = r^2$. We called it an *implicit function*. We never said explicitly what $y$ was as a function of $x$, but we could tell from the equation what values of $y$ went with what values of $x$."

"All we need to do, then, is figure out how to take the derivative of an implicit function," I said. Slowly I began to see a vision of a chain surrounding the implicit function $y^q = x^p$. "We can use the chain rule," I suggested. "Suppose we have $y = (g(x))^2$ and we want to find $dy/dx$."

"We can use the chain rule," the king said. "We have $f(u) = u^2$, and $u = g(x)$. It looks like $dy/dx = (dy/du)(du/dx) = 2u(du/dx)$."

"How does that help?" the professor demanded. "Our problem is $y^q = x^p$."

"We can use what we just did to find the derivative of both sides of the equation," I said.

"Both sides?" Recordis asked. "How can you do that?"

"All the equation says is that we have two functions—one of them is $y^q$ and the other one is $x^p$—and that they're equal for all $x$. That means that they're really two different ways of expressing the same function. And that means that their derivatives with respect to $x$ should be equal. Therefore we can apply the operator $d/dx$ to both sides."

$$\frac{d}{dx}\, y^q = \frac{d}{dx}\, x^p$$

"The part on the right is easy," Recordis said, "since we know that $p$ is an integer."

$$\frac{d}{dx}\, x^p = px^{p-1}$$

$$\frac{d}{dx}\, y^q = px^{p-1}$$

"How can you take the derivative of a function if it has a $y$ in it?" Recordis demanded.

"We can use what we just did," the king said. "We said that if (some function) $= y^q$, then $(d/dx)$ (some function) $= qy^{q-1}(dy/dx)$."

"We don't know what $dy/dx$ is," the professor objected. "That is what we are trying to solve for."

"But we can solve for it now," the king said. "Our original equation becomes as follows."

$$qy^{q-1}\frac{dy}{dx} = px^{p-1}$$

Recordis looked at the equation suspiciously, as if he wasn't sure it was a legitimate one. "You could try dividing both sides by $qy^{q-1}$," he suggested.

$$\frac{dy}{dx} = \frac{px^{p-1}}{qy^{q-1}}$$

"We can substitute $y = x^{p/q}$," the king noted.

$$\frac{dy}{dx} = \frac{p}{q}\frac{x^{p-1}}{(x^{p/q})^{q-1}}$$

"We can simplify that exponent in the denominator," the king noticed. (Remember that $(x^a)^b = x^{ab}$.)

$$\frac{dy}{dx} = \frac{p}{q} \frac{x^{p-1}}{x^{p-p/q}}$$

"We can use another exponent law to simplify the fraction with the $x$ in it," Recordis noted. (Remember that $x^a/x^b = x^{a-b}$.)

$$\frac{dy}{dx} = \frac{p}{q} x^{(p-1)-(p-p/q)}$$

"That can be simplified," the professor noted.

$$\frac{dy}{dx} = \frac{p}{q} x^{p-1-p+p/q}$$

$$= \frac{p}{q} x^{-1+p/q}$$

$$= \frac{p}{q} x^{p/q-1}$$

"That's what we wanted to show!" Recordis cried happily. "That means that, if $y = x^n$, $y' = nx^{n-1}$, whether or not $n$ is an integer."

"We should make up a name for this rule, since it seems to be true in general," the king said.

"That's easy," the professor announced. "We'll call it the *power rule*."

---

**POWER RULE**

$$y = f(x) = x^n \qquad (n \text{ is any rational number})$$

$$y' = f'(x) = nx^{n-1}$$

---

(We later established that the power rule is valid for any real number exponent, including irrational numbers, such as $\pi$ or $\sqrt{2}$.)

"We should save the method we thought of to prove this," Recordis said. "The implicit function method."

Igor wasn't sure how to generalize this, so I thought of a way to start.

"The problem will come when we have an expression like this: $f(y) = g(x)$. If we take the derivative with respect to $x$ of both sides, $(d/dx) f(y) = (d/dx) g(x)$, we know that the chain rule tells us that $(df/dy)(dy/dx) = dg/dx$. Now we can solve for $dy/dx$."

Igor displayed the function for which we could now allegedly find the derivative:

$$A = w(r^2 - w^2)^{1/2}$$

"First we should use the product rule," the professor said. "We would have $dA/dw$ equals the first function $(w)$ times the derivative of the messy function $((r^2 - w^2)^{1/2})$ plus the derivative of $w$ (which is 1) times the messy function."

$$\frac{dA}{dw} = (w)\,\frac{d}{dw}\,(r^2 - w^2)^{1/2} + (1)(r^2 - w^2)^{1/2}$$

"Now all we have to do is find the derivative of the messy function," the king noted. "I think we'll need both the chain rule and the power rule."

$$\frac{d}{dw}\,(r^2 - w^2)^{1/2} = \tfrac{1}{2}(r^2 - w^2)^{(1/2)-1}\,\frac{d}{dw}\,(r^2 - w^2)$$

"I know what the derivative of $(r^2 - w^2)$ is," Recordis said. "The $r^2$ is a constant, so its derivative is zero. The derivative of $-w^2$ is $-2w$."

$$\frac{d}{dw}\,(r^2 - w^2)^{1/2} = \tfrac{1}{2}(r^2 - w^2)^{-1/2}(-2w)$$

$$= -w(r^2 - w^2)^{-1/2}$$

"We can put that expression back in our equation for $dA/dw$," the king said.

$$\frac{dA}{dw} = -w^2(r^2 - w^2)^{-1/2} + (r^2 - w^2)^{1/2}$$

"Now we have to do what we did yesterday," the professor pointed out. "We set $dA/dw$ equal to zero and then solve for the optimal value of $w$."

$$0 = -w^2(r^2 - w^2)^{-1/2} + (r^2 - w^2)^{1/2}$$

"Now we can do it," Recordis said. "That's just algebra."

$$w^2(r^2 - w^2)^{-1/2} = (r^2 - w^2)^{1/2}$$
$$w^2(r^2 - w^2)^{-1/2}(r^2 - w^2)^{1/2} = (r^2 - w^2)^{1/2}(r^2 - w^2)^{1/2}$$
$$w^2 = (r^2 - w^2)$$
$$2w^2 = r^2$$
$$w^2 = \frac{r^2}{2}$$
$$w = \frac{r}{\sqrt{2}}$$

"That's the answer!" Recordis shouted. "I need to design my house so that the width of the living room is equal to the radius of the circle divided by the square root of 2!"

"I remember what the radius of your yard is," the professor said. "It's 10 units. So that means that the width should be as follows." (Recordis spent a long time looking up the square root of 2 in a table and performing the division.)

$$w = (10)(2)^{-1/2} = \frac{10}{1.1414} = 7.07$$

"Shouldn't we check the second derivative?" the king asked. "It would be embarrassing if this turned out to be a minimum point and Recordis ended up building the smallest possible house."

"I think we can just tell if we look at some points on the original function," the professor said. "That would save us from having to check the second derivative."

Igor made a table of values of the function $A(w) = w(100 - w^2)^{1/2}$ (Table 4–1).

"Our answer is right!" Recordis said. "The maximum value of the area is somewhere about 7. It is amazing the uses we are finding for this subject of calculus. And I bet with these three new rules we can find the derivative of almost anything." Recordis was still talking cheerfully about plans for his house when we all went out to breakfast.

| Table 4–1 | |
|---|---|
| $w$ | $A$ |
| 0 | 0 |
| 1 | 9.95 |
| 2 | 19.60 |
| 3 | 28.62 |
| 4 | 36.66 |
| 5 | 43.30 |
| 6 | 48.00 |
| 7 | 49.99 |
| 8 | 48.00 |
| 9 | 39.23 |
| 10 | 0 |

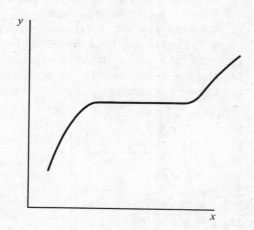

Figure 4–6.

## Note to Chapter 4

The assumption that has been made in the demonstration of the chain rule in this chapter is that $\Delta u$ does not equal zero. Since $\Delta u = g(x + \Delta x) - g(x)$, this condition means that $g(x + \Delta x)$ must not equal $g(x)$ for small values of $\Delta x$. This condition holds true for any function without any flat spots. For example, this demonstration of the chain rule will not work for the function in Figure 4–6. We later were able to establish, however, that the chain rule is indeed valid for all functions, whether or not they have any flat spots.

## Exercises

1. Use the product rule to find the derivative of the function $f(t) = (2t - 5)(3t + 4)$. (Does the answer agree with the result for exercise 2–7?)

Find $dy/dx$ for:

2. $y = (2x + 5)/(3x - 1)$
3. $y = 10x/(4x - 3)$
4. $y = (ax + b)/(cx + d)$
5. Use the product rule to find the derivative of $y = (x^2)(x^3)$.
6. Derive the triple product rule (i.e., find $y'$ for $y = u(x) \, v(x) \, w(x)$). Use the triple product rule to find the derivative of $y = x^3$.
7. Derive the quotient rule. (Find $y'$ for $y = u(x)/v(x)$.)
8. Find $dy/du$ for $y = u^{3/2}$. Find $du/dx$ for $u = x^2 + 3$. Then find $dy/dx$ for $y = (x^2 + 3)^{3/2}$.
9. Consider $y = \sqrt{1 + x^2}$. Let $u = 1 + x^2$. Find $dy/du$. Find $du/dx$. Find $dy/dx$.
10. Consider $y = (3 + 4x)^{3/2}$. Let $u = 3 + 4x$. What is $dy/du$? What is $du/dx$? What is $dy/dx$?
11. Consider $y = \sqrt{ax^2 + bx + c}$. Let $u = ax^2 + bx + c$. What is $dy/du$? What is $du/dx$? What is $dy/dx$?
12. Consider $y = \sqrt{1 + (x^2 + 4)^{-1}}$. Let $u = x^2 + 4$. Let $v = 1 + (x^2 + 4)^{-1}$. Find $dy/dv$. Find $dv/du$. Find $du/dx$. Find $dy/dx$.
13. Verify the power rule for $n = \frac{1}{2}$. Use the definition of the derivative.
14. Use the definition of the derivative to find $y'$ for $y = 1/x$.
15. Derive a generalized power rule: for $u = u(x)$, find $dy/dx$ for $y = u^n$.

Find $dy/dx$ for:

16. $y = \sqrt{4 + 2x}$
17. $y = (4 + 3x)^{3.3}$
18. $y = \sqrt{3x^3 + 2x^2 + x}$
19. $y = 3/\sqrt{x^2 - 1}$
20. $y = \sqrt[3]{x}$
21. $y = \sqrt[3]{x^2 + 4}$
22. $y = \sqrt{1 + 1/x}$
23. $y = \sqrt{x + 1/x}$
24. $y = (x^2 + 4)^{-1}$
25. Find the second derivative of the function $A(w) = w\sqrt{r^2 - w^2}$ at the point where $w = r/\sqrt{2}$. Is the curve concave up or concave down at that point? Is Recordis building a house that will have maximum area or a house that will have minimum area?
26. (a) Use the implicit derivative method to find $dy/dx$ for the circle $x^2 + y^2 = r^2$. (b) Solve for $y$ as an explicit function of $x$ for the semicircle where $y$ is positive. (c) Using the chain rule and the power rule, find $dy/dx$ for the function in (b). Does the result agree with that obtained in (a)?
27. Find $dy/dx$ for the hyperbola $(y - 1)^2/9 - (x - 3)^2/4 = 1$.

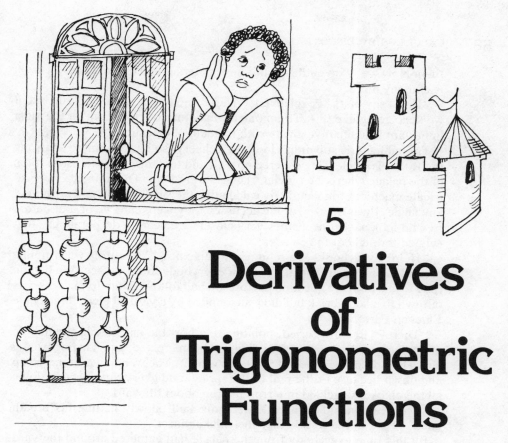

# 5
# Derivatives of Trigonometric Functions

In the next few days I saw more of the pretty countryside of Carmorra. Recordis had announced he was starting construction of his house, and he promised to invite us all over for a visit when it was finished.

While I was walking along the balcony of the castle the next afternoon, I noticed Trigonometeris leaning against the railing. He apparently hadn't heard me the first time I called to him, so I became concerned and ran up to him. "Are you all right?" I asked.

He jumped and looked around at me. "Oh," he said. "Sure." His voice lacked its usual cheerful quality.

"Something's the matter," I insisted. "Are you sick?"

"Nothing really," he said. "Just a feeling of uselessness."

"Uselessness?"

"I suppose I can't stand in the way of progress," he said. "I guess this is what happens when new things are invented—the old inventors get left behind. I always wondered what happened to the inventors of drag-sleds after the wheel was invented, and now I know."

"What are you talking about?" I asked.

"My work was crucial when we were working on trigonometry," he said. "When we were discovering the new trigonometric identities, I was called upon to reach the limits of my skill. Trigonometry was the in-thing then. Now we're inventing all this new material, and trig just doesn't seem useful any more.''

''Don't say that!'' I exclaimed. ''There's no telling what we'll run into if we keep going like this. Trigonometry may turn out to be one of the most important fields to investigate with the help of calculus.''

Trigonometeris still didn't look very cheery, so I asked him to show me the Royal Triangles. He cheered up a bit and took me to the treasure room of the palace where the official triangles were kept. The triangles were intricate machines for accurately calculating the sine, cosine, and tangent of any angle. I was amazed at the accuracy they were capable of, but before I could look at the mechanics very closely we heard a desperate cry that echoed throughout the palace.

''Help! Everybody! We're in real trouble now!'' Recordis was running around, trying to get the attention of everybody in the palace. We had no time to ask him any questions before we had followed him to the outskirts of town to a large brick building surrounded by a yard labeled ''McCockle Chicken Farm.''

''There!'' Recordis cried, pointing to a horrible machine in the middle of the farmyard. A giant scaffolding had been erected, and hanging from it was a heavy spring attached to a large weight. The weight was jumping up and down because of the pull of the spring, and every time it came down all the chickens cackled in terror and ran about the yard.

''The gremlin put it there!'' Recordis said, slowly getting his breath back. ''He's trying to sabotage the chicken farm!''

By this time everybody from the palace had gathered around the yard (except Mongol, who was terrified by the spring and was hiding behind a nearby mountain). ''The gremlin put locks all around the spring so we can't take it apart,'' Recordis added.

''We've got to find some way to stop it,'' the king said. ''Before long the chickens will have been scattered all over the place.''

Builder, whom I had not seen since the day I arrived, looked thoughtfully at the block. ''I could build a gremlin-block-stopping machine,'' he said. ''All I need to know is how fast the block is moving at a given time.''

''We can take a derivative!'' the professor cried. ''That will tell us how fast it is going.''

''Take a derivative of what?'' Recordis asked. ''We can't tell how fast it is going before we know what a graph of its motion looks like.''

''It's very simple,'' the king said. ''We need to find some way to make a graph of its motion, then we find a function of time that matches the graph, and then we find the derivative of that function to find the speed of the block.''

''I don't know any function that goes up and down like that!'' Recordis protested. ''And, besides, if you think I'm going to climb up on that block with my watch to time its position . . .''

''There must be another way to graph its motion,'' the professor said.

"We could use a camera," I suggested. I explained my idea, and we all worked quickly to set the camera up. Mr. McCockle was shouting at us to hurry because the chickens were quite hysterical by now.

"When the pictures are developed we will make life-sized prints of them, and then we can measure the position of the block," the professor said. For some reason there is very little friction in Carmorra, and the spring was still bouncing up and down at exactly the same rate while we took the pictures, had them developed, and laid them out on the floor of the Main Conference Room. (See Figure 5–1.)

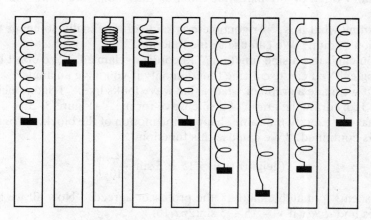

**Figure 5–1.**

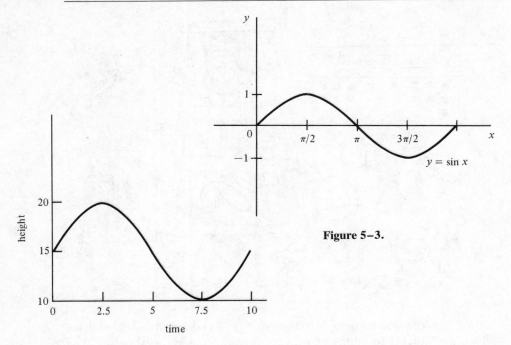

**Figure 5–3.**

**Figure 5–2.**

"Make sure you put them in the right order," the professor advised. "It would surely confuse matters if you didn't."

"That looks familiar," Trigonometeris murmured, but nobody paid him any attention.

"Let's have Igor draw the graph of a function that fits these points," the professor said. (Figure 5–2.)

"I can't think of any algebraic function that looks like that," the king stated.

"I know what it is," Trigonometeris said softly. Finally someone paid attention to him. "We can use a sine function."

"You can't use a sine function!" Recordis exclaimed. "We don't have any angles! You can use a sine function only if you have an angle."

"But we made a graph of what a sine wave looks like," Trigonometeris said. "Igor, draw the graph of a sine wave for me." (Figure 5–3.)

"That has exactly the same shape as the motion of the block," Trigonometeris continued. "We can use this function."

$$\text{(height)} = y = 15 + 5 \sin\left(\frac{2\pi t}{10}\right)$$

"It seems to match exactly," the professor agreed. "Now all we have to do is find $dy/dx$ if $y = 15 + 5 \sin(2\pi t/10)$."

"Let's first find the derivative of $y = \sin x$," I suggested. "Then we can easily use the chain rule, the product rule, and the sum rule to find the derivative of the actual function that we have.''

"We might as well start at the beginning," the professor said. "Igor, show us the definition of the derivative again."

$$\frac{dy}{dx} = \lim_{\Delta x \to 0} \frac{f(x + \Delta x) - f(x)}{\Delta x}$$

If $f(x) = \sin x$ then

$$\frac{dy}{dx} = \lim_{\Delta x \to 0} \frac{\sin(x + \Delta x) - \sin x}{\Delta x}$$

"Now we're stuck," Recordis said.

"I have a formula for the sine of the sum of two angles," Trigonometeris informed us triumphantly. He went to his book and read off the following formula:

$$\sin(A + B) = \sin A \cos B + \cos A \sin B$$

"We can put that formula in the equation for the derivative," Trig suggested.

$$\frac{dy}{dx} = \lim_{\Delta x \to 0} \frac{\sin x \cos \Delta x + \cos x \sin \Delta x - \sin x}{\Delta x}$$

"And we can separate the terms in the numerator," he added.

$$\frac{dy}{dx} = \lim_{\Delta x \to 0} \sin x \left( \frac{\cos \Delta x - 1}{\Delta x} \right) + \lim_{\Delta x \to 0} \cos x \frac{\sin \Delta x}{\Delta x}$$

$$\frac{dy}{dx} = \sin x \lim_{\Delta x \to 0} \frac{\cos \Delta x - 1}{\Delta x} + \cos x \lim_{\Delta x \to 0} \frac{\sin \Delta x}{\Delta x}$$

"Now we have to figure out what these limits are," Trigonometeris said. "Let's try this one: $\lim_{\theta \to 0} (\sin \theta)/\theta$."

"I think the limit will be zero," the professor said; "$\sin \theta$ will approach zero, and that is on the top of the fraction. Whenever the top of the fraction goes to zero, the value of the fraction is zero."

"I think it will be infinity," Recordis objected. "Remember that we have a $\theta$ on the bottom. Whenever the bottom of the fraction goes to zero, the value of the fraction goes to infinity."

"Are we measuring $\theta$ in radian or degree measure?" the king asked.

"We had better use radian measure," Trigonometeris answered. "If I remember correctly, we found that degree measure was more convenient if we were actually building something or measuring a real angle, but for mathematical purposes we always used radian measure."

"Then I think the value of the limit will be 1," the king said. "Igor, draw a picture of a circle with radius 1." (Figure 5–4.)

"Remember that this dotted line is equal to sin $\theta$, and this deep black curve is equal to $\theta$. It looks as though they come closer and closer together as $\theta$ approaches zero."

"Fascinating," the professor said. "I bet you can't prove it, though."

I ventured a suggestion. "Suppose we draw another line." (Figure 5–5.)

"It looks as though the shaded area is smaller than the striped area, which is smaller than the area of the whole big triangle."

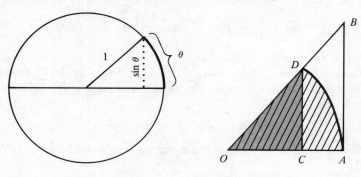

**Figure 5–4.**                    **Figure 5–5.**

"We can write that as an inequality," the professor said.

$$(\text{area shaded}) < (\text{area striped}) < (\text{total area})$$

"Two of those areas are triangles, and one area is a sector of a circle," the professor went on. "We should be able to figure out what all those areas are. First, we'll label all the points. Then the area of the dotted triangle is $\frac{1}{2}(O - C)(C - D)$, and the area of the whole triangle is $\frac{1}{2}(O - A)(A - B)$."

"We derived a formula for the area of a sector of a circle," Recordis said.

$$(\text{area of sector of circle}) = \tfrac{1}{2}\theta r^2$$

where $\theta$ is the angle enclosed by the sector and $r$ is the radius of the circle.

"So that would make the striped area equal to $\frac{1}{2}\theta(1)^2 = \frac{1}{2}\theta$."

We then rewrote the inequality:

$$\tfrac{1}{2}(O - C)(C - D) < \tfrac{1}{2}\theta < \tfrac{1}{2}(O - A)(A - B)$$

"Now we can substitute trigonometric values in the expression," Trig said.

$$(O - C) = \cos \theta$$

$$(C - D) = \sin \theta$$

$$(O - A) = 1$$

$$(A - B) = \tan \theta$$

Igor displayed the rewritten inequality:

$$\tfrac{1}{2}\cos\theta\sin\theta < \tfrac{1}{2}\theta < \tfrac{1}{2}\tan\theta$$

"We can multiply everything by 2," the king noted. "That's legal with an inequality, as long as we multiply by a positive number."

$$\cos\theta\sin\theta < \theta < \tan\theta$$

"Remember that $\tan\theta = \sin\theta/\cos\theta$," Trigonometeris said.

$$\cos\theta\sin\theta < \theta < \frac{\sin\theta}{\cos\theta}$$

"We can divide everything by $\sin\theta$," the king suggested.

$$\cos\theta < \frac{\theta}{\sin\theta} < \frac{1}{\cos\theta}$$

"Now we can find the limit!" Trigonometeris said. "As $\theta$ goes to 0, $\cos\theta$ goes to 1. That means $1/\cos\theta$ goes to 1, also. And that means that $\lim_{\theta\to0}(\theta/\sin\theta)$ gets squeezed in between two numbers that are both going to 1!"

"That must be a tight fit," Recordis remarked.

"And if $\theta/\sin\theta$ approaches 1, then $\sin\theta/\theta$ must approach 1, too. That means the king was right."

$$\lim_{\theta\to0}\frac{\sin\theta}{\theta} = 1$$

Igor displayed the formula we had for the derivative of the sine function:

$$\frac{dy}{dx} = \sin x \lim_{\Delta x\to0}\frac{\cos\Delta x - 1}{\Delta x} + \cos x(1)$$

"Now we're stuck," Recordis mourned.

"No, all we have to do is evaluate that other limit," the king said.

That turned out to be tedious but not especially difficult, so I just wrote down the steps we took.

$$\lim_{\theta\to0}\frac{1-\cos\theta}{\theta} = \lim_{\theta\to0}\frac{1-\cos\theta}{\theta}\cdot\frac{1+\cos\theta}{1+\cos\theta}$$

$$= \lim_{\theta\to0}\frac{1-\cos^2\theta}{\theta(1+\cos\theta)}$$

$$= \lim_{\theta\to0}\frac{\sin^2\theta}{\theta(1+\cos\theta)}$$

$$= \lim_{\theta\to0}\frac{\sin\theta}{\theta}\lim_{\theta\to0}\frac{\sin\theta}{1+\cos\theta}$$

$$\lim_{\theta\to0}\frac{1-\cos\theta}{\theta} = \lim_{\theta\to0}\frac{\sin\theta}{1+\cos\theta}$$

As $\theta$ goes to 0, cos $\theta$ goes to 1:

$$\lim_{\theta \to 0} \frac{1 - \cos \theta}{\theta} = \lim_{\theta \to 0} \frac{\sin \theta}{2}$$

$$= 0$$

"Now we can solve the problem!" Trigonometeris rejoiced.

$$y = \sin x$$

$$\frac{dy}{dx} = \sin x \lim_{\Delta x \to 0} \frac{\cos \Delta x - 1}{\Delta x} + \cos x \lim_{\Delta x \to 0} \frac{\sin \Delta x}{\Delta x}$$

$$= \cos x$$

"Amazing!" the king exclaimed. "Does it make sense?"

"When $x = 0$ the curve is sloping gently upward," Trigonometeris said. "Then cos $x = 1$, so the slope of the curve should be 1. That looks reasonable."

"At $x = \pi/2$ the sine curve has a horizontal tangent," the professor said.

"And, sure enough, cos$(\pi/2)$ is zero!" the king said.

We checked a few more points along the curve, and this surprisingly simple relationship seemed to check out all along the curve.

"I wonder what the derivative of cos $x$ is," the king wondered.

"It's sin $x$, I bet," Recordis said.

"Let's check that," the king said. (Figure 5–6.)

"The slope of the curve can't be sin $x$," the professor objected. "You can tell that just by looking at these few points."

"It could be $-\sin x$," Recordis guessed.

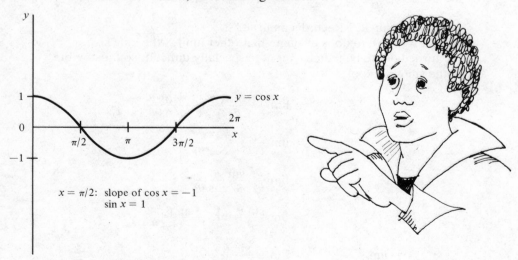

$x = \pi/2$:  slope of cos $x = -1$
          sin $x = 1$

**Figure 5–6.**

"I know a quick way to check," Trigonometeris said. "Instead of trying to find the derivative of $y = \cos x$ directly, we could take advantage of the fact that $\cos x = \sin(\pi/2 - x)$. With the chain rule we should be able to find the derivative of that function easily."

$$y = \cos x = \sin\left(\frac{\pi}{2} - x\right)$$

Let $u = \pi/2 - x$. Then

$$y = \sin u$$

$$\frac{dy}{dx} = \frac{dy}{du}\frac{du}{dx}$$

"We know that $du/dx = -1$," Recordis said.
"And $dy/du = \cos u$, since that is what we just did," the king added.

$$\frac{dy}{dx} = (\cos u)(-1) = -\cos\left(\frac{\pi}{2} - x\right) = -\sin x$$

"Recordis, you were right!" the professor exclaimed.
"I was? I mean—of course, I was right." Recordis beamed happily.
We made a table of these rules:

---

## DERIVATIVES OF TRIGONOMETRIC FUNCTIONS

| | |
|---|---|
| $y = \sin x$ | $dy/dx = \cos x$ |
| $y = \sin u$ | $dy/dx = \cos u\,(du/dx)$ |
| $y = \cos x$ | $dy/dx = -\sin x$ |
| $y = \cos u$ | $dy/dx = -\sin u\,(du/dx)$ |

---

"Now we can solve the original problem and save those poor chickens," Trigonometeris said.

$$y = 15 + 5\sin\left(\frac{2\pi t}{10}\right)$$

$$\frac{dy}{dt} = \pi\cos\left(\frac{2\pi t}{10}\right)$$

Recordis quickly wrote the answer down and handed it to Mongol, who had gotten over his fear of the spring. He ran to find Builder so Builder could begin work on the block-stopping machine.

"Let's find the derivative of $y = \tan x$," Trigonometeris said. "Then we can have a complete table."

"We can use the product rule," the professor pointed out.

$$y = \tan x$$

$$= (\sin x)(\cos x)^{-1}$$

$$\frac{dy}{dx} = \sin x \frac{d}{dx}(\cos x)^{-1} + (\cos x)^{-1}\frac{d}{dx}\sin x$$

$$= \sin x \frac{d}{dx}(\cos x)^{-1} + \frac{\cos x}{\cos x}$$

$$= (\sin x)(-1)(\cos x)^{-2}(-\sin x) + 1$$

$$\frac{dy}{dx} = \frac{\sin^2 x}{\cos^2 x} + 1$$

"That's not as complicated as I thought it would be," Recordis said.

"We can rewrite it, using $\tan x = (\sin x)/(\cos x)$," Trigonometeris told him.

$$\frac{dy}{dx} = \tan^2 x + 1$$

"I have another one," Trigonometeris said. "We derived this trigonometric identity."

$$\tan^2 x + 1 = \sec^2 x$$

We rewrote the derivative:

$$y = \tan x$$

$$\frac{dy}{dx} = \sec^2 x$$

We decided to find the derivatives for the rest of the trigonometric functions. For cotangent we merely inserted $\mathrm{ctn}\, x = \tan(\pi/2 - x)$, and came up with $dy/dx = -\csc^2 x$. We puzzled over the derivative of the secant function for a minute before the professor realized we had already solved that problem since $\sec x = (\cos x)^{-1}$. We decided to write the answer as:

$$y = \sec x$$

$$y' = \tan x \sec x$$

Igor displayed the results of our work for the day:

| | |
|---|---|
| $y = \sin u$ | $dy/dx = (\cos u)\, du/dx$ |
| $y = \cos u$ | $dy/dx = (-\sin u)\, du/dx$ |
| $y = \tan u$ | $dy/dx = (\sec^2 u)\, du/dx$ |
| $y = \mathrm{ctn}\, u$ | $dy/dx = (-\csc^2 u)\, du/dx$ |
| $y = \sec u$ | $dy/dx = (\tan u \sec u)\, du/dx$ |
| $y = \csc u$ | $dy/dx = -(\mathrm{ctn}\, u \csc u)\, du/dx$ |

We decided to return to the McCockle Chicken Farm and watch Builder destroy the gremlin's terrible machine. As we were walking out of the room, Trigonometeris started talking excitedly. "I never realized how fascinating this whole subject—this calculus—could be. We can make some use of these springs as toys for children, provided, of course, that we make them small enough so they won't scare any chickens. And now we can make sine-curve-shaped slides, and we will know what the slope is at any point! I'll have to start designing a new playground right away!"

By the time we reached the farm we found that Builder had indeed neutralized the deadly spring, and the chickens were all happily gathered about Mr. McCockle. And I was glad that this little adventure had provided new life for the career of someone as kind and polite as Alexanderman Trigonometeris.

_____ Exercises

Find the derivatives of the following functions:

**1.** $y = \sin x^2$

**2.** $y = \sin^2 x$

**3.** $y = 1/(\sin x)$

**4.** $y = x \sin x$

**5.** $y = (\sin x)(\cos x)$

**6.** (a) Use the formula for $\sin(A + B)$ to find $y'$ for $y = \sin(x^2 + x)$.

(b) Use the chain rule to find the derivative of the function in (a).

7. If $A$ is an angle expressed in degree measure, find $y'$ for $y = \sin A$.

8. Find the values of $x$ where $y = \sin x$ has horizontal tangents. Which of these points are minima? Which are maxima? In what intervals is the curve concave upward? Concave downward?

9. Find the acceleration of the block in the gremlin's machine: $y = 15 + 5 \sin(2\pi t/10)$.

10. Trigonometeris knows that $\sin 30° = \sin(\pi/6) = \frac{1}{2}$. Find the equation of the tangent line to the curve $y = \sin x$ at the point $(\pi/6, \frac{1}{2})$. What is the radian equivalent for 32°? Use the tangent line derived above to find an approximate value for $\sin 32°$.

11. Use the power rule and the chain rule to find the derivative $(dy/d\theta)$ of $y = (1 + \tan^2 \theta)^{1/2}$. What is a simpler way to express this function?

12. Consider the curve $y = \tan \theta$ in the interval $\theta = -\pi/2$ to $\theta = \pi/2$. Where does this curve have horizontal tangents? Where is it concave up? Where is it concave down?

13. Calculate the value of the function $f(x) = (\sin x)/x$ for each of the following values of $x$: 0.785, 0.5, 0.3, 0.1, 0.05.

14. A block with mass $m = 2$ kg attached to a spring behaves according to the equation $-kx = m(d^2x/dt^2)$, where $k$ is known as the *spring constant* for this particular spring (measured in kilogram-meters per second squared). The motion of the block is given by $x = 0.8 \sin 3t$. Find the value of $k$.

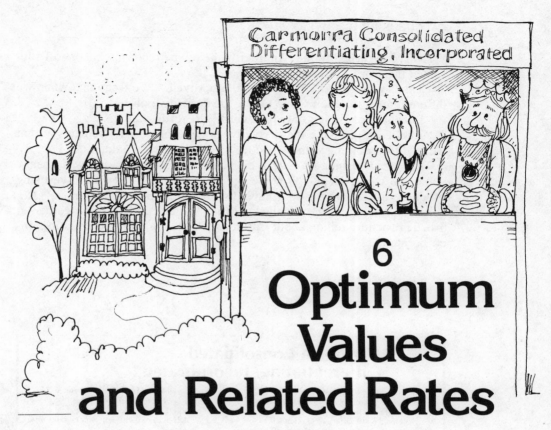

# 6
# Optimum Values and Related Rates

The king had a new concern when we met in the Main Conference Room a few days later. "We need to start planning some way for us to make the results of everything we've done available to the people. It doesn't seem right that we should use calculus only to solve our own problems."

"I know what we can do!" the professor said. "We can make lots of money with these discoveries!"

"That's it!" Recordis cried. "We'll start a company that will solve calculus problems for people! I bet there are lots of calculus problems floating around that people haven't recognized before because they haven't known what calculus is. If we charge people to solve these problems, we should be able to make a nice profit," he said, with dollar signs flashing in his eyes.

"Let's think of a catchy name for our company," the professor said before the king had a chance to protest. "How about Carmorra Consolidated Derivative-Taking Business?"

"We need a shorter name," Recordis objected. "For instance, when we take the derivative of something, we have to say we're taking the derivative of it. That's three words! We need one word that stands for the process of taking a derivative."

Everyone turned to me. Another name popped into my head, so I suggested the word *differentiate*.

"That sounds good," the professor approved. "We'll say that when we differentiate a function we're taking the derivative of it."

"I like it because it's only one word," Recordis said.

"And if we're taking the derivative with respect to *x*, we can say that we're differentiating with respect to *x*," the professor added.

They went rushing on with their plans, although the king was uneasy about the whole idea. "Now we need to have Igor draw an advertising brochure for us, which we'll have distributed all over the country," Recordis continued. A short while later they had helped Igor put together an artistic brochure telling about Carmorra Consolidated Differentiating, Incorporated.

## *Carmorra Consolidated Differentiating, Incorporated

This unique new firm uses all the mysterious powers of the newly discovered subject of calculus to provide public services involving a wide variety of problems that can occur in any possible application! We offer many services to our customers:

- We can find the slope of the tangent line for any curve, no matter how complex.
- We can calculate the speed or acceleration (if we're given the position as a function of time) of any conceivable object.
- With the aid of the derivative we can determine the properties (such as critical points or maximum points) of any possible curve (including the amazing trigonometric functions). Just try to determine the properties of a fifth-degree polynomial without consulting calculus!
- For any function we can find all local maxima and minima and all absolute maxima and minima. We can thus solve a wide variety of problems that occur in construction, economics, geometry, and many other fields.
- Special rules allow calculations for derivatives of functions made up of any number of added functions, multiplied functions, or composite functions.
- The unique chain rule allows the calculation of two different related rates of change.
- Special tables are available for trigonometric functions (which allow us to handle all kinds of problems involving oscillating motion).
- We also provide translation into mathematical language; this means that you don't even need to supply us with explicit descriptions of the functions you need! We have a wide variety of modern methods, such as photographic devices, to allow us to turn a set of data that you provide into a function.

We're anxious to serve the calculus needs of the public. We have pretabulated lists available for standard problems, and we offer custom service for unusually complex problems. Check our rates, and we're sure you'll find that they are less than those of any other consolidated differentiating firm in all of Carmorra.

HOURS FOR BUSINESS: 10:00 A.M. to 5:00 P.M.
Monday to Wednesday
ADDRESS: 10 Central Avenue, Capital City

Our brochures were printed and distributed all over the country. Builder constructed a small stand just outside the palace with a prominent sign bearing the company name and emblem. We worked out a schedule so we could trade off our various times for running the company, but at first we were so interested that we all crowded around awaiting our first customer.

"Remember the list of prices I made up," the professor whispered.

At first people weren't used to the business, and the first day there were no customers. We were all getting nervous during the second day, but in the afternoon our first customer finally showed up.

"I hear you folks can solve optimum-value problems," the man drawled.

("I know him!" Recordis whispered to me. "He's the man who runs the company where I buy my supply boxes for my pens and pencils.")

"Yes, we can," the professor said, glad that our first customer had arrived while it was her turn to be in charge of the business.

"I need to make a box," the box-maker said. "I have 2 square meters of wood to use. The box needs to have square ends and an open top. (See Figure 6–1.) Can you tell me the dimensions of the box that will allow me to get the maximum possible volume for my 2 square meters of material?"

"Sure," the professor answered. "First, we'll set up the variable names we need. We'll let $x$ equal the length of one of the square edges, and $y$ equal the length of one rectangular edge. Then the volume will be (volume) $= V = x^2y$. We can also figure out the surface area, which will be the area of the two square ends plus the area of the three rectangular sides: (surface area) $= A = 2x^2 + 3xy$."

The professor was getting nervous, so the king helped her figure out what to do next. "We know that the surface area equals 2, since that is what you told us."

$$2 = 2x^2 + 3xy$$

"We can solve for $y$ algebraically!" Recordis said.

$$2 - 2x^2 = 3xy$$

$$y = \frac{2}{3x} - \frac{2x}{3}$$

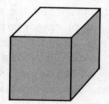

"We can put that expression for $y$ back into the equation for the volume," the king pointed out.

**Figure 6–1.**

$$V = x^2y = x^2\left(\frac{2}{3x} - \frac{2x}{3}\right) = \frac{2x}{3} - \frac{2x^3}{3}$$

"Now we have exactly what we want," the king continued. "The volume is expressed as a function of one variable (in this case $x$). All we need to do is find the derivative and set it equal to zero to solve for the optimum value of $x$."

$$V = \frac{2x}{3} - \frac{2x^3}{3}$$

$$\frac{dV}{dx} = \tfrac{2}{3} - 2x^2$$

$$0 = \tfrac{2}{3} - 2x^2$$

$$\tfrac{2}{3} = 2x^2$$

$$x^2 = \tfrac{1}{3}$$

$$x = \frac{1}{\sqrt{3}}$$

"Now we can easily figure out $y$," the professor said, regaining her composure. After some calculations, she told the customer, "You should make your box with edges 0.577 meter by 0.577 meter by 0.769 meter. That will give you a total volume of 2.56 liters."

"Not bad," the box-maker said. "A box that big will hold all I need."

"Now about the charge . . ." The professor started looking through her rate sheet for a typical optimum-value problem.

"Hold it!" Recordis said. "We can't charge him! He has always been very generous in providing me with the boxes that I keep my pens and pencils in, so the least we can do for him is to solve his optimum-value problem without charge."

The professor wanted to protest. The king agreed with Recordis, however, so the box-maker went home with a free solution to his optimum-value problem.

"We do need a paying customer now," Recordis said.

A while later an elderly gentleman carrying a large stack of magazines approached the stand. ("That's the publisher of *Carmorra Magazine!*" the professor gasped.)

"I understand that if I give you some information you can tell me how to maximize something, correct?"

"Of course," said Recordis, since it was his turn to run the business.

"I need to maximize the profits of the firm I run. I need to decide what price to charge for a subscription to *Carmorra Magazine*. If the price is lower, more people will subscribe."

"That sounds reasonable," Recordis agreed.

"I did some calculations and made some measurements to determine how many magazines I would be able to sell at a particular price. It turns out that the number of subscriptions will be as follows."

(number of subscriptions) $= n = -5000p + 15,000$

"For example, if the price is \$2.50 I will sell 2500 subscriptions."

"Then you should sell for a very low price, and you will have lots and lots of subscribers," Recordis suggested.

"But I also have to pay something for each magazine I publish," the customer went on. "It costs me \$1.65 to print each magazine, so if I lower the price too much I will end up losing money. Can you figure out the price that maximizes my profits?"

"In principle it's pretty easy," Recordis said. "We need two variables —the variable to be maximized and the variable that can be adjusted to the optimum value. We never did anything like this with magazine subscriptions, though."

"I know what we have to do," the king said. "If $Y$ is the amount of profit, then we need to write $Y$ as a function of $p$."

"A what of what?" the publisher asked.

"And profit is given by (profit) $= Y =$ (revenue) $-$ (cost)."

"We know that cost is $1.65n$," the professor noted.

"And revenue must be equal to $pn$," Recordis added. "But we don't know what $p$ and $n$ are, and we can't solve a problem with two variables in it."

"We can use the relationship between $p$ and $n$," the king said.

$$n = -5000p + 15,000$$

We put that expression into the equation for revenue:

$$(\text{revenue}) = pn = p(-5000p + 15,000) = -5000p^2 + 15,000p$$

"Now we can put that expression back into the equation for profit!" Recordis said.

$$Y = -5000p^2 + 15,000p - (1.65)(-5000p + 15,000)$$

$$= -5000p^2 + 15,000p + (1.65)5000p - (1.65)(15,000)$$

$$= -5000p^2 + 15,000p + 8250p - 24,750$$

$$Y = -5000p^2 + 23,250p - 24,750$$

"Now it's easy!" Recordis said. "All we need to do is take the derivative and set it equal to zero."

$$\frac{dY}{dp} = -10,000p + 23,250$$

$$0 = -10,000p + 23,250$$

$$10,000p = 23,250$$

$$p = \frac{23,250}{10,000}$$

$$= 2.32$$

"And that's the optimum price!" Recordis said. "Now about the charge . . ."

The magazine publisher was stunned to see such a mysterious number pop out of the calculus hocus-pocus so quickly. "Hold it," he said. "Can you calculate what my profits would be if I set the price at $2.32?"

"Sure," the king said. "First we calculate how many magazines you will sell."

$$n = (-5000)(2.32) + 15,000 = 3400$$

"Then we calculate total revenue."

$$(\text{revenue}) = pn = (2.32)(3400) = 7888$$

"And we figure total cost."

$$(\text{cost}) = (1.65)(3400) = 5610$$

"So you would have the following profit."

$$(\text{profit}) = 7888 - 5610 = 2278$$

"Not bad," the publisher commented, impressed.

"I like your magazine," the king said. "I like all the pretty pictures of the streams and mountains in Carmorra."

"That's partly why we need these profits. We want to make sure that the streams and mountains stay unspoiled and unpolluted. I hate to seem mistrustful, but before I set the subscription price at $2.32 I'd like to be sure that this really is the best price and that I can't do better with a price of $2.20 or $2.40 or something else."

"Oh, that's easy," Recordis said. He turned to a page in his notes and was about to explain the whole theory of derivatives and horizontal tangents. The king could see that the publisher wasn't interested, and thought of a better idea.

"Let's calculate your profits for a price of $2.20 and of $2.40 and see what happens." He went through the same calculations (Table 6–1).

**Table 6–1**

| | | |
|---|---|---|
| Price | $2.20 | $2.40 |
| Number sold | 4000 | 3000 |
| Total revenue | $8800 | $7200 |
| Total cost | $6600 | $4950 |
| Profit | $2200 | $2250 |

Recordis was nervous during the calculations, but when they were finished he cried out triumphantly, "See! If you make the price $2.40 or $2.30, you will make less profit than you will if you set the price at $2.32. Now about the charge . . ."

By now the publisher was deeply impressed, but the professor interrupted him before he could hand over any money. "We can't charge him! He needs these profits to make sure the streams and mountains stay pretty. Also, he has been very generous to the king and the Royal Court in his editorials in the past.'' She didn't add that *Carmorra Magazine* had once done a large feature on herself and that she was hoping the staff would do another one sometime soon.

"We'll let you have this answer free,'' the king said. The publisher walked away to plan the next issue of the magazine before Recordis had a chance to protest.

"We need some paying customers now,'' the professor noted.

"I hope we get some other kinds of problems,'' Recordis said. "Not that I don't like maxima/minima problems, but I would prefer a little variety.''

A while later a young mother came up to our desk, and it was the professor's turn to wait on her.

("I recognize her!'' Recordis whispered. "She lives only a couple of houses away from me. She has such a nice family.'')

"I understand you solve problems involving related rates of change,'' she said. "I'm planning a surprise birthday party for one of the children in the neighborhood. I worked out a very nice set of balloons that I would like to inflate suddenly at the moment of the surprise. I designed and built a special Variable Rate Air Pumper to inflate the balloons. I would like to have the balloons inflate so that their radii are increasing at constant rates. Can you tell me at what rate I should pump air into each balloon to make this happen?''

"Shouldn't you pump air into the balloon at a constant rate?'' the professor asked. "If $v$ represents the volume of the balloon and $t$ represents time, then we would say in our notation that $dv/dt = c$, where $c$ is some constant.''

"That won't work because the balloon is getting bigger!'' the customer said. "As the balloon gets bigger, we need to pump more air into it per given interval of time in order to keep the radius increasing at a constant rate.''

"We need the chain rule, don't we?'' the king suggested. "That rule tells us how to deal with related rates.''

"If you say so,'' the professor said. "I always say that when in doubt you should write down what you know. We know that $dr/dt = 1$, where $r$ is the radius. We also know from geometry that $v = (4/3)\pi r^3$. And we want $dv/dt$ as a function of time. Now what?''

"We can find $dv/dr$,'' the king told her helpfully; "$dv/dr = 4\pi r^2$.''

"I see how we can get $dv/dt$ using the chain rule,'' the professor said suddenly.

$$\frac{dv}{dt} = \frac{dv}{dr}\frac{dr}{dt}$$

$$= (1)(4\pi r^2) = 4\pi r^2$$

"That's the answer!'' Recordis said.

"Hold it!'' the professor objected. "We need to find $dv/dt$ as a function of time, but we have it as a function of the radius. Now we need to find out what $r$ is as a function of time.''

"We should be able to figure that out, since we know that the radius is increasing at the rate of 1 centimeter per second,'' the king said. "How big is the balloon when you start blowing air into it?'' he asked the young mother.

"One centimeter in radius,'' she said.

"Then $r$ must be equal to $t + 1$,'' the king said. We all agreed that this function had the desired properties: $r$ was 1 when $t$ was 0 and $r$ increased at a constant rate of 1 centimeter per second.

"That gives us the answer for how much air to put in at a given time,'' the professor said.

$$\frac{dv}{dt} = 4\pi(1 + t)^2$$

"Thank you very much,'' the customer said. "I had guessed that the answer would involve the second power of time.''

"Now about the charge . . .'' the professor began.

"We can't charge someone who's planning a birthday party for some children!'' Recordis cried. "How could you be so heartless! We'll let you have this answer free.'' The young mother went home grateful for her neighbor Recordis.

"We must have some paying customers now,'' Recordis said. "Otherwise we won't be able to stay in business.''

A few minutes before closing time a young man came up to our stand outside the palace.

("I know him!'' the professor whispered. "He's the lifeguard at National Park Beach!'')

"I understand you solve problems related to the speed of things. I'd like to figure out how fast my shadow moves,'' he said. "At night I turn on a light at the top of my lifeguard station. When I start running away from the tower at a constant speed, my shadow runs ahead of me. Can you figure out how fast the shadow is moving?''

"Certainly,'' Recordis said. "First, we'll call the height of the light $h$ and we'll call your height $L$ (for lifeguard). We'll call the distance you've moved from the tower $x$.

"In the notation we use, the speed at which you're running is represented by $dx/dt$. We'll call that constant speed $v$.''

$$\frac{dx}{dt} = v$$

"We may as well call the distance from the shadow to the tower $s$.'' (Figure 6–2.)

"Now all we have to do is find $ds/dt$,'' he continued. "Which we do by . . .''

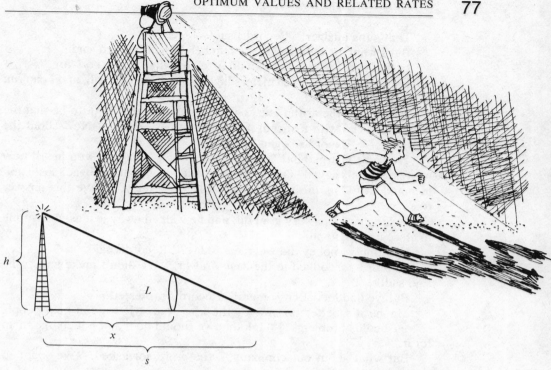

**Figure 6–2.**

"Using the chain rule," the king interrupted. "We can use a similar triangle relationship to tell us that $h/s = L/(s - x)$."

"That's easy," the professor said. "We know that corresponding sides of similar triangles are in proportion."

We found $dx/ds$:

$$s - x = \frac{sL}{h}$$

$$x = s - \frac{sL}{h}$$

$$\frac{dx}{ds} = 1 - \frac{L}{h}$$

The king showed us how we could use the chain rule:

$$\frac{ds}{dt} = \frac{dx/dt}{dx/ds}$$

We inserted the expressions we had found:

$$\frac{ds}{dt} = \frac{v}{1 - L/h}$$

"That's the answer!" Recordis said.

"But I want an answer that's a number," the lifeguard said.

"In that case, all you have to do is tell us what $v$, $L$, and $h$ are."

"My lifeguard tower is 8 meters high, I'm 1.8 meters high, and I can run 6.7 meters per second."

Recordis did the arithmetic. The final answer turned out to be that the shadow moved with a speed of 8.6 meters per second. "Now about the charge . . ." Recordis began.

"We can't charge him!" the professor said. "Haven't you heard how many lives he's saved? One of my best friends was at the beach last summer and owes his life to this lifeguard. We'll let you have this answer free."

The lifeguard was very grateful, and he walked away just as it was time to close up our stand.

"How much money did we make today?" Recordis said.

The professor looked in the cash drawer. "We didn't make any," she said sadly.

"But we had lots of customers!" Recordis protested.

"Maybe it's better this way," the king said. "If we can help people solve calculus problems, I think that we should do it and not charge them for it."

"But what about our company?" the professor asked. "We spent so much time working on the brochures."

"We'll make it a nonprofit company," the king said. "That way we will still stay in business answering people's questions, but we won't need to worry about how much money we make."

We all decided to accept that plan, so before we returned to the palace we made a slight change in our sign so that it read:

# Carmorra Consolidated Differentiating, Incorporated
A NONPROFIT AGENCY DESIGNED TO SERVE YOU!

## Exercises

1. The conical Central Park reservoir is $a$ units across and $b$ units deep. Water is flowing into it at the rate of $u$ cubic units per minute. How fast is the surface of the water rising when the water is $h$ units deep? Suppose $a = 3$ meters, $b = 6$ meters, and $u = 0.2$ cubic meter per minute. What is $dh/dt$ when $h = 3$ meters?

2. One day Builder left a ladder of length $L$ leaning against the side of the palace, and it started to slide. The bottom of the ladder slid along the ground with a constant speed of $u$ meters per second. How fast was the top of the ladder sliding down the wall?

3. Find the point on the curve $y^2 = 8x$ that is closest to the point (4, 2).

4. If a rectangle has a fixed perimeter equal to $k$, what is the shape of the rectangle that will have the maximum area?

5. If a rectangle has a fixed area of $A$, what shape will minimize the perimeter?

6. The profit function of a perfectly competitive firm is given by $Y = Px - C(x)$. $P$, the price of output, is a fixed number beyond the control of the firm. $C(x)$ is the cost function, which measures the cost of producing $x$ units of output. The marginal cost ($MC$) is the derivative of the cost function. Find a condition, stated in terms of $MC$ and $P$, that the firm must meet if it is to maximize its profits.

7. Recordis' special boxes have volume $V$ and height $h$. The material to be used for the base and sides of the box costs \$$r$ per square unit, and the material to be used for the lid costs \$$2r$ per square unit. Find the dimensions of the box that minimize the total cost of the box.

8. Recordis has another kind of box with the top, three sides, and the base made of the material that costs \$$r$ per square unit, and the front side (dimensions $x$ by $h$) made of the material that costs \$$2r$ per square unit. Find the shape of the box that minimizes the cost.

9. Find the shape of the right circular cylinder that has volume $V$ and the minimum possible total surface area.

10. At what point is the slope of the curve $y = x^4 - 3x^3 + 5x - 10$ the greatest?

11. Recordis releases his toy boat from South Beach, and it travels due north at 5 cm/sec. At the same time, Mongol releases his toy boat from East Beach, which is $20\sqrt{2}$ meters due northeast from South Beach. Mongol's boat travels due west at 7 cm/sec. How far apart are the two boats at time $t$? At what time are they closest together? How close together are they at the moment that they are closest together?

12. Mongol has a rope $L$ units long that is strung over a pulley $h$ units high. One end is attached to a large weight on the ground. Mongol holds the other end and starts walking away from the pulley at a constant speed of $u$ units per second. (When he starts walking the rope is taut.) How fast does the weight move up?

13. A car is to be driven on a trip $D$ units long. The amount of fuel that the car uses per hour is given by $10v^2 - 100v + 290$, where $v$ is the speed of the car in units per hour. The car will travel with a constant speed throughout the trip. What should $v$ be so that the total fuel consumption for the trip will be minimized?

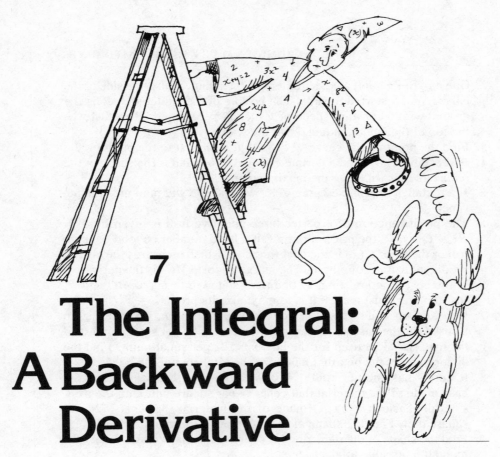

# 7

# The Integral: A Backward Derivative

The differentiation business went along smoothly. The professor told everyone that calculus was the most enjoyable invention since the Hasselbluff Mountain Sky Sled, and she was confident that we had discovered everything there was to know about mathematics.

Every day we held a meeting in the Main Conference Room. Slowly we began to notice that each day Recordis arrived at the meeting out of breath. A few days after the formation of the company he was panting so hard that the professor finally asked him what was happening.

"It's Rutherford, my dog," Recordis said. "Lately he's been very frisky, and I have to run all over the yard to catch him."

"Where does he run?" the king asked.

"He always runs in a straight line, because that's where he buried his bones. He treats that stretch of ground like a racetrack. He changes his speed, though, so I can never figure out where he is. I chase him all morning until he finally gets tired and I can take him back inside the house. He knocks down the neighbors' flowers if I let him run around loose when I'm not there."

"Can't you figure out a better way to catch him?" the professor asked.

"I can stand on a ladder and drop his collar down on him," Recordis said. "That only works, though, if I know exactly where he will be at a given time, so I know the exact moment to drop the collar. I try to guess the right moment, but I usually guess wrong. He comes by quicker than I'm expecting, or slower than I'm expecting, so that the collar falls just in front of him or just behind him. I do know how fast he goes, which might be useful, but I can't figure out from that what his position will be.''

"How do you know how fast he goes?" the professor asked.

"I can tell because I feed him scientifically designed dogfood," Recordis answered. "It's designed so that it releases energy to Rutherford at a certain rate. For example, if I feed him Special Speedup Wonder Dogfood he will be running faster all the time. His speed at time $t$ will be equal to $v = 4t$. Or I can feed him Patented Wearout Wonder Dogfood, so that he runs very fast right at the beginning but tires quickly and loses speed. Then his speed is equal to $v = 60 - 4t$. If only I knew his position! If, for example, I knew that 10 seconds after he started running he would be 200 units from his can of dogfood, I could plan to drop the collar so that it would land there exactly at $t = 10$ seconds. I shouldn't bother you with my problems, though.

"I wish we could do something for you," the professor said. "It isn't good to have a record keeper who is out of breath every day. Still, we should get back to the business of differentiating."

"Wait a minute," the king interrupted. "Couldn't we do it backwards?"

"Do what backwards?" the professor asked.

"Differentiation. We know that, if we start with a position function and differentiate it, we end up with a velocity function. If we start with a velocity function and differentiate backwards, we should end up with a position function."

There was a long silence.

"How would you differentiate something backwards?" Recordis asked. "I have enough trouble differentiating things forwards."

"We should be able to figure out some way to do it," the king said. "For example, let's take Recordis' problem with Rutherford. When he eats the speedup dogfood, we know that his velocity is $v = dx/dt = 4t$. All we have to do is think of some function whose derivative is $4t$."

"Try $t^2$," the professor guessed.

"That won't work," Recordis said. "We know that $(d/dt)t^2 = 2t$, which isn't equal to $4t$."

"Try $t^3$," Trigonometeris suggested.

"No," the professor said. "We know that $(d/dt)t^3 = 3t^2$."

"$4t^2$?" Recordis guessed.

"No," Trigonometeris said. "We know that $(d/dt)4t^2 = 8t$."

"I know what will work," the king said slowly. "We know that the answer must have a $t^2$ in it, because the derivative will have a $t^1$ in it. Try $2t^2$."

Igor did the differentiation:

$$x = 2t^2$$

$$\frac{dx}{dt} = 4t$$

"It works!" the professor cried.

"Maybe I can outwit that dog," Recordis said. "We must have a name for a backward derivative. I'm not going to write 'backward derivative' all the time."

"We could call it an *antiderivative*," the professor said. We agreed that that was one possible name.

$$\text{position function} = \text{antiderivative} = x = 2t^2$$

$$\text{velocity function} = \frac{dx}{dt} = 4t$$

"That sounds too destructive," Trigonometeris said. "We should think of a constructive sounding name, too."

Everybody turned to me, and an interesting name suddenly popped into my mind. "We'll call it an *integral*," I said.

"That sounds impressive," Recordis agreed.

$$\text{position function} = \text{antiderivative} = \text{integral} = x = 2t^2$$

$$\text{velocity function} = \frac{dx}{dt} = 4t$$

"We must think of a scientific way to calculate integrals, or antiderivatives," the king said. "Guessing won't work most of the time. How can we know ahead of time that the answer is $2t^2$ and not $100t^3$ or $2t^2 + 99$ or something else?"

"We shall have to make a set of rules," Recordis answered. "We wrote a set of rules for calculating derivatives, so we should be able to develop a set of rules for calculating integrals."

"Hold everything!" the professor objected. "We have done something horribly wrong! How *do* we know that the answer is not $2t^2 + 99$?"

"We check the derivative," the king said.

$$x = 2t^2 + 99$$

$$\frac{dx}{dt} = \frac{d}{dt} 2t^2 + \frac{d}{dt} 99 = 4t + 0 = 4t$$

The king stopped short.

"That position function works, too!" Recordis cried. "They both work! However, one of them must be wrong, because I know that I have only one dog and he can be in only one place at one time. There is no way that two different functions can describe his position."

"You can't ever compute an integral," the professor said. "Suppose you know that $dx/dt = 4t$. That will be the case if $x = 2t^2 + 7$ or $x = 2t^2 + 86.234567$ or, in fact, if $x = 2t^2 + C$, where $C$ is any constant number.

That is hopelessly indefinite.''

"We'll have to call it an *indefinite integral*," Recordis said.

"We'll have to forget about this and go back to our regular business," the professor stated firmly.

"I'm sorry I ever brought the subject up," Recordis said.

The professor was about to go on with the day's business when the king came over to me and whispered something in my ear. "I think you're right," I whispered back. "The king has an idea," I told everybody. "Igor, draw a graph of several curves that obey the relationships $dx/dt = 4t$, $x = 2t^2 + C$." (See Figure 7–1.)

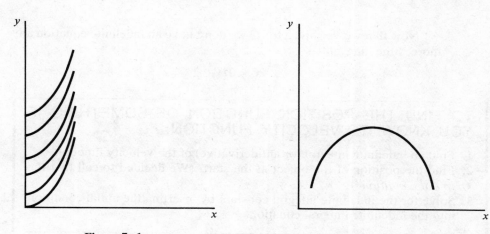

Figure 7–1.                              Figure 7–2.

"They all look like the same curve," the king said. "The only difference is that they have been moved up and down."

"That's why the idea of an integral should be useful," I explained. "I don't think it is hopelessly indefinite. For example, we know that a function with a different shape won't work (Figure 7–2) because its derivative isn't equal to $2t$. The only curves that will work are the curves on the first graph. It is useful to know that much."

"If we knew which one of these curves is the right one, then we would know exactly where Rutherford is all the time," the king said. "How are we going to tell which curve is the right one?"

We stared hard at the collection of curves on Igor's screen. We were looking for a clue that would make the single guilty curve stand out as the culprit responsible for Rutherford's devious motion.

"The difference between the curves seems to be where Rutherford starts from," the king said. "For example, on the first curve it looks as though Rutherford starts from position zero when $t = 0$. For the other curves, he starts farther along than the zero mark."

"Then it should be easy to tell which curve is the right one!" the professor exclaimed. "Recordis, where is Rutherford when he starts running?"

"I put up a little stake to measure the distance along the track from the house. He starts at his can of dogfood, which is right next to stake number 2."

"That means $x = 2$ when $t = 0$," the professor stated. "That will tell us which curve is the right one!"

"We need to solve for the arbitrary constant $C$," the king said. "Let's put in the values $x = 2$ when $t = 0$."

$$x = 2t^2 + C$$

$$2 = 2(0)^2 + C$$

$$C = 2$$

"Now that we've solved for $C$ we don't have an indefinite equation any more," the king added.

$$x = 2t^2 + 2$$

---

## TO FIND THE POSITION FUNCTION OF SOMETHING IF YOU KNOW ITS VELOCITY FUNCTION:

1. Find an indefinite integral (or antiderivative) of the velocity function.
2. Find the position of the object at the start. (We decided to call that the *initial condition*.)
3. Solve for the indefinite integral constant by inserting the initial condition into the indefinite integral equation.

---

"We need a symbol for an integral," Recordis said.

"When we have a function $f(x)$ we call its derivative $f'(x)$," the professor offered. "Suppose we call its antiderivative $F(x)$."

"Using a capital letter!" Recordis said. "Ingenious! We need another symbol, too, to correspond to the $dx/dt$ notation."

"We can figure something out," I said. "We start with $dx/dt = 4t$, and we need some way to turn that $dx/dt$ into just plain $x$."

"We can get rid of the $dt$ by multiplying both sides by $dt$," Trigonometeris suggested.

"No, you can't," the professor objected. "That would work if $dx/dt$ were a fraction, because then it would be $dx$ divided by $dt$. But it's not a fraction. It's a derivative."

"Derivatives do seem to behave a little bit like fractions," the king said. "Remember the chain rule: $dy/dx = (dy/du)(du/dx)$."

"We can make a definition of what $dx$ means," I said. "Igor, draw a picture of a curve with its tangent line (Figure 7–3). Let's let $dx$ equal this deep black line, and $dt$ equal this dotted line. Then $dx$ divided by $dt$ will equal the slope of the tangent line, which is what we know it must be."

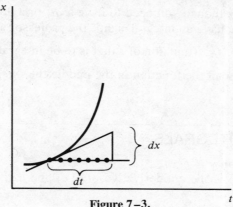

**Figure 7–3.**

We made the definition:

$$dx = \frac{dx}{dt}\, dt$$

"Still, $dx$ is not a number that you can specify a value for," the professor pointed out. "We just said that it was some small number. We do need to think of a name for it. Since these numbers come from differentiating a function, we could call them *differentials: dx* means differential $x$, and $dt$ means differential $t$." (We later established that the most important thing to remember about differentials is that a differential does not mean anything if it is all by itself. For example, the equation $dx = 4t$ does not mean anything. Differentials must always come in pairs, as in $dx = 4t\, dt$, or in a derivative, as in $dx/dt = 4t$, or with an integral sign, as in $x = \int f(t)\, dt$.)

We rewrote our equation for the derivative using differential notation:

$$\frac{dx}{dt} = 4t$$

$$dx = 4t\, dt$$

"How do we get rid of the $d$ in front of the $x$?" Recordis asked.

"We'll cross it out," I said cheerfully.

$$\int dx = 4t\, dt$$

"Hold it!" the professor said. "That's an equation. If you do something to one side of the equation you must do the same thing to the other side!"

"All right," I said. "We'll do it like this."

$$\int dx = x = \int 4t\, dt$$

"Pretty!" Mongol muttered.

"If he likes it we had better keep it," Recordis said. "We'll call $\int$ the *integral sign*."

"It looks as though you need to have a *dt,* or a differential something, whenever you have an integral sign," the professor said.

$$x = \int (\text{function of } t \text{ that is to be integrated}) \, dt$$

"We could call the function in the middle the *integrand,*" the king remarked.

---

## INDEFINITE INTEGRALS

Start with a function $y = f(x)$.
The function $F(x)$ is the antiderivative of $f(x)$ and satisfies the following condition:

$$\frac{dF(x)}{dx} = f(x)$$

The symbol $\int f(x) \, dx$ represents the indefinite integral of $f(x)$, and it is equal to $F(x) + C$, where $F(x)$ is the antiderivative and $C$ is any constant number (it is known as the *arbitrary constant of integration*).
The constant $C$ can be determined if you know an initial condition.

---

"We'll have to figure out lots of indefinite integrals," Recordis said. "Rutherford has several different kinds of dogfood, and he'll stop eating Special Speedup Wonder Dogfood if he knows I've figured out his position function. For example, what should we do when his speed is given by $dx/dt = 60 - 4t$?"

First we multiplied by *dt:*

$$dx = (60 - 4t) \, dt$$

Then we wrote the integral sign on both sides of the equation:

$$\int dx = \int (60 - 4t) \, dt$$
$$x = \int (60 - 4t) \, dt$$

"It would be useful if we could find a rule for the integral of the sum of two functions," the professor said. "Suppose we have the following."

$$y = \int (f(x) + g(x)) \, dx$$

"I bet that's equal to this," Recordis said.

$$y = \int f(x) \, dx + \int g(x) \, dx$$

"Why do you guess that?" the professor asked.
"That's the way it worked for derivatives," Recordis answered.
"We'll take the derivative of your guess and see if it is equal to $f(x) + g(x)$," the professor said.

$$y = \int f(x) \, dx + \int g(x) \, dx$$

$$\frac{dy}{dx} = \frac{d}{dx} \int f(x) \, dx + \frac{d}{dx} \int g(x) \, dx$$

$$= \frac{d}{dx} (F(x) + C) + \frac{d}{dx} (G(x) + C)$$

$$= \frac{d}{dx} F(x) + \frac{d}{dx} C + \frac{d}{dx} G(x) + \frac{d}{dx} C$$

$$= f(x) + 0 + g(x) + 0$$

$$\frac{dy}{dx} = f(x) + g(x)$$

"See! I was right again," Recordis said. "We'll call that the *sum rule for integrals*."

---

## SUM RULE FOR INTEGRALS

$$\int (f(x) + g(x)) \, dx = \int f(x) \, dx + \int g(x) \, dx$$

---

Using the sum rule, we rewrote the integral we were trying to evaluate:

$$x = \int 60 \, dt - \int 4t \, dt$$

"It would be nice if we could figure out what to do with numbers like 60 when they occur inside an integral sign," Recordis said. "I wish we could pull that 60 outside the integral sign, so we could write $60 \int dt$ instead of $\int 60 \, dt$. Integrals are nice, but I think the fewer numbers we have in the middle of them the better."

We tried Recordis' suggested rule. We wanted a function whose derivative was $n \, f(x)$, and Recordis' guess was that we should try $n \int f(x) \, dx$.

$$\frac{d}{dx} n \int f(x) \, dx = n \frac{d}{dx} \int f(x) \, dx$$

(We know that constants move across derivatives.)

$$\frac{d}{dx} n \int f(x) \, dx = n \frac{d}{dx} (F(x) + C)$$

$$= n(f(x) + 0)$$

$$= n \, f(x)$$

"It does work!" Recordis said. "We'll call that the *multiplication rule for integrals*."

---

## MULTIPLICATION RULE FOR INTEGRALS

$$n \int f(x)\ dx = \int n\, f(x)\ dx$$

(where $n$ is any constant number)

---

"Let's extend that rule to the case where we have a variable number," Recordis went on eagerly. "I bet we can say that

$$\int q(t)\, f(t)\ dt = q(t) \int f(t)\ dt$$

where $q$ is any variable number."

Igor took the derivative of $q(t) \int f(t)\ dt$ to see whether it was equal to $q(t)\, f(t)$, as Recordis' theory said it would be.

$$x = q(t) \int f(t)\ dt$$

$$\frac{dx}{dt} = \frac{d}{dt}\, q(t) \int f(t)\ dt$$

From the product rule:

$$\frac{dx}{dt} = q(t)\, \frac{d}{dt} \int f(t)\ dt + \frac{dq(t)}{dt} \int f(t)\ dt$$

$$= q(t)\, f(t) + \frac{dq(t)}{dt} \int f(t)\ dt \neq q(t)\, f(t)$$

"That doesn't work!" the king exclaimed.

"I was wrong," Recordis said apologetically.

The professor was irritated. "We'll have to add an amendment to the multiplication rule. It looks as though the integral sign acts as a filter. You can move *constants* across integral signs as much as you want, but you can never move *variables* across an integral sign."

---

## MULTIPLICATION RULE FOR INTEGRALS

$$\int n\, f(t)\ dt = n \int f(t)\ dt$$

(where $n$ is a constant number)

However, if $n$ is a variable, then it may *not* be taken outside the integral sign, and $\int n\, f(t)\ dt \neq n \int f(t)\ dt$.

---

Using the multiplication rule, we rewrote the problem for the wearout dogfood.

$$\frac{dx}{dt} = 60 - 4t$$

$$x = 60 \int dt - 4 \int t\, dt$$

(Since $t$ is a variable, it had to stay inside the integral sign.)

"We know what $\int dt$ is," Recordis said. "That's just equal to $(t + C)$."

"How do you know that?" the professor asked.

"Because $\int dt$ is really the same thing as $\int (1)\, dt$, which is the same as $dx/dt = 1$. That means we need to find a function whose derivative is 1. The function that satisfies that condition is 1 times $t$, but of course we need to add that constant thingamajig to it."

We decided to make that a rule. Sometime later we agreed to call an integral of the form $\int dt$ a *perfect integral*, so we called the rule the *perfect integral rule*.

---

### PERFECT INTEGRAL RULE
$$\int dt = t + C$$

---

Igor rewrote our equation:

$$x = 60t + C + 4 \int t\, dt$$

"Now we have to take care of that $t$ in the integral sign," the king said. We thought about this problem for a long time.

Finally the professor had an idea. "What we need to do is work the power rule backwards. The power rule says that, if $y = x^n$, then $dy/dx = nx^{n-1}$. Now suppose we have $dy/dx = x^n$, and we need to figure out what $y$ is."

"I bet it will have an $x^{n+1}$ in it," the king said. "If we take the derivative of $x^{n+1}$, we get $(n + 1)x^n$."

"But we don't want $(n + 1)x^n$," the professor objected. "We want to end up with just plain old $x^n$."

"It would work if we could find some way to get rid of the $(n + 1)$," the king said.

"I know how we can get rid of it," Recordis contributed. "Try this."

$$y = \frac{1}{n + 1} x^{n+1}$$

"Are you sure that will work?" the professor said, amazed that Recordis was coming up with so many answers in the same day.

$$y = \frac{1}{n+1} x^{n+1}$$

$$\frac{dy}{dx} = \frac{d}{dx} \frac{1}{n+1} x^{n+1}$$

$$= \frac{1}{n+1} \frac{d}{dx} x^{n+1}$$

$$= \frac{1}{n+1} (n+1) x^n$$

$$\frac{dy}{dx} = x^n$$

"It does work!" the king said. We called this rule the *power rule for integrals*.

---

## POWER RULE FOR INTEGRALS

$$\int x^n \, dx = \frac{1}{n+1} x^{n+1} + C$$

---

Now we easily solved for Rutherford's *position* when he ate the wear-out dogfood:

$$x = \int (60 - 4t) \, dt$$

$$= \int 60 \, dt - 4\int t^1 \, dt$$

$$= 60t - (4)(\tfrac{1}{2} t^2) + C$$

$$x = 60t - 2t^2 + C$$

Using the initial condition $x = 2$ when $t = 0$, we got

$$2 = (60)(0) - 2(0)^2 + C$$

$$C = 2$$

$$x = 60t - 2t^2 + 2$$

"I hate to cause more problems," Recordis said, "but I just thought of another kind of dogfood that Rutherford has. What do we do if we have something other than $x$ raised to a power? For example, what if we have $\int u^n \, dx$, where $u$ is a function of $x$? Rutherford has one kind of dogfood where his speed is given by $dx/dt = \sqrt{3t + 5}$. In this case we have $\int u^{1/2} \, dt$, where $u = 3t + 5$."

"We need to find some way to turn that $dt$ into a $du$," the king told him. "Then we can use the power rule."

"I bet we can do that," the professor noted. "We have the derivative: $du/dt = 3$. We can write that in differential notation: $du = 3dt$."

"That means we can rewrite the integral!" the king said.

$$x = \int \sqrt{3t + 5} \left(\tfrac{1}{3}\right) 3dt = \int u^{1/2} \tfrac{1}{3} du$$

"We can pull that $\tfrac{1}{3}$ outside the integral, since it is just a constant," Recordis said.

$$x = \tfrac{1}{3} \int u^{1/2} du$$

"Now we can use the power rule!" the professor exclaimed.

$$x = \left(\tfrac{1}{3}\right) \left(\tfrac{2}{3}\right) u^{3/2} + C$$

$$= \tfrac{2}{9} u^{3/2} + C$$

"But we want an answer in terms of $t$, not in terms of $u$," Recordis protested.

"We can always make the reverse substitution," the king said. "Since $u = 3t + 5$, we have $x = \tfrac{2}{9}(3t + 5)^{3/2} + C$."

We called this method the *method of integration by substitution.*

---

## INTEGRATION BY SUBSTITUTION

If you face an integral of the form $\int u^n \, dx$, where $u$ is a function of $x$, you cannot use the power rule directly until you have converted the $dx$ into a $du$. First, find the derivative of $u$ ($du/dx$) and write that derivative using differential notation. Then make the substitution $dx = (dx/du) \, du$, so the integral becomes equal to

$$\int u^n \, dx = \int u^n \frac{dx}{du} \, du$$

If $dx/du$ is a constant, it can be moved outside the integral sign:

$$\int u^n \, dx = \left(\frac{dx}{du}\right) \int u^n \, du = \frac{dx}{du} \frac{1}{n+1} u^{n+1} + C$$

(If $dx/du$ is not a constant, as in $\int \sqrt{1 - x^2} \, dx$, you can tell that we will be in real trouble. We worry about that kind of problem in Chapter 11.)

---

"Does Rutherford have any cans of dogfood that make his speed a trigonometric function?" Trigonometeris asked, trying to be useful. "If he does, it should be no problem, since we can easily say that $\int \cos x \, dx = \sin x$, and $\int \sin x \, dx = -\cos x$."

"Now I better hurry home and catch him!" Recordis said, ignoring Trigonometeris. We made a table of our results. We had found everything we needed to know to integrate polynomials, so naturally Recordis thought that we had done all we would ever need to do with integrals.

## INDEFINITE INTEGRALS

Start with a function $f(x)$.
$F(x)$ is a function such that $dF/dx = f(x)$, and it is called the antiderivative of $f(x)$.
$\int f(x)\, dx$ is the indefinite integral of $f(x)$ and is equal to $F(x) + C$.
The value of the arbitrary constant $C$ can be determined if you know the initial condition.

## SUM RULE FOR INTEGRALS

$$\int (f(x) + g(x))\, dx = \int f(x)\, dx + \int g(x)\, dx$$

## MULTIPLICATION RULE FOR INTEGRALS

$$\int n\, f(x)\, dx = n \int f(x)\, dx$$

if $n$ is a constant, but not if $n$ is variable.

## PERFECT INTEGRAL RULE

$$\int dx = x + C$$

## POWER RULE FOR INTEGRALS

$$\int x^n\, dx = \frac{1}{n + 1}\, x^{n+1} + C$$

## Exercises

Evaluate the following. (Remember that the answer to an indefinite integral is a function plus an arbitrary constant.)

1. $y = \int (x^2 + 3x + 5)\, dx$
2. $y = \int (ax^2 + bx + c)\, dx$
3. $y = \int (9x + 10)\, dx$
4. $y = \int (14)\, dx$
5. $y = \int (x^3 + 1)\, dx$
6. $y = \int (6x^5 + 10x^3 + 3x)\, dx$
7. $y = \int (4x^3 + 3x^2 + 2x + 1)\, dx$
8. $y = \int \left( \dfrac{x^3}{3} + \dfrac{x^2}{2} + x \right) dx$

9. $y = \int (x^m + x^n)\, dx$

10. $y = \int [(m + 1)x^m + (n + 1)x^n + (p + 1)x^p]\, dx$

11. $y = \int x^{100}\, dx$

12. $y = \int \sin \theta\, d\theta$

13. $y = \int \cos \theta\, d\theta$

14. $y = \int \sec^2 \theta\, d\theta$

15. $y = \int \left( \dfrac{1}{x^4} + \dfrac{1}{x^3} \right) dx$

16. $y = \int (x^2 + x^{-2})\, dx$

17. $y = - \int \csc^2 \theta\, d\theta$

18. $y = \int \sec \theta \tan \theta\, d\theta$

19. $y = -\int \csc \theta \operatorname{ctn} \theta\, d\theta$

20. $\int \frac{1}{2}(1 - \cos 2\theta)\, d\theta$

21. $\int \sin^2 \theta\, d\theta$

22. Evaluate $x \int x^2\, dx$ and $\int xx^2\, dx$.

Find $x(t)$ for each of the following situations. Think of an example of an object that might move according to each equation.

23. $dx/dt = 0$ $\qquad\qquad$ $t = 0,\ x = 5$

24. $dx/dt = 4$ $\qquad\qquad$ $t = 0,\ x = 2$

25. $dx/dt = 55$ $\qquad\quad$ $t = 8,\ x = 175$

26. $dx/dt = at$ $\qquad\quad$ $t = 0,\ x = 0$

27. $dx/dt = at$ $\qquad\quad$ $t = 0,\ x = x_0$

28. $dx/dt = \cos at$ $\qquad$ $t = \pi/2a,\ x = \frac{1}{2}$

29. $dx/dt = t^4$ $\qquad\qquad$ $t = 1,\ x = 1$

30. $d^2x/dt^2 = a$ $\qquad$ $t = 0,\ x = x_0;\ t = 0,\ dx/dt = v_0$

31. $d^2x/dt^2 = 0$ $\qquad$ $t = 0,\ x = x_0;\ t = 0,\ dx/dt = v_0$

Differentiate each of these functions and express the answer using differential (rather than derivative) notation:

32. $y = x^2$

33. $y = \sin x$

34. $y = cx$

35. $y = x^{1/2}$

36. $y = (1 + x^2)^{1/2}$

Solve the following integrals. Use the substitution indicated to convert the integral into a form where the power rule or some other simple rule can be used.

37. $y = \int x\sqrt{1 - x^2}\, dx;$ $\qquad$ let $u = 1 - x^2$

38. $y = \int x^2 \sqrt{a + bx^3}\, dx;$ $\qquad$ let $u = a + bx^3$

39. $y = \int x \sin(x^2)\, dx;$ $\qquad$ let $u = x^2$

40. $y = \int \sin^3 x \cos x\, dx;$ $\qquad$ let $u = \sin x$

Evaluate:

41. $y = \int \dfrac{x}{(5 + 6x^2)^2}\, dx$

42. $y = \int x^{n-1} \sqrt{a + x^n} \, dx$
43. $y = \int \sec^2 x \tan^3 x \, dx$
44. $y = \int x^9 \sin(x^{10}) \, dx$
45. $y = \int u^n \, (du/dx) \, dx$
46. $y = \int \sin \theta \, (d\theta/dx) \, dx$
47. When Mongol throws his beach ball into the air, its acceleration is given by $dv/dt = -g$, with the initial condition $v = v_0$ when $t = 0$. (a) Find the velocity of the ball at time $t$. (b) Find the height of the ball at time $t$, using the initial condition $h = 0$ when $t = 0$.
48. When Mongol drops his ball off Hasselbluff Mountain, its acceleration is given by $d^2h/dt^2 = -g$, with the initial conditions $h = 64$ and $dh/dt = 0$ when $t = 0$. Find the velocity of the ball at time $t$. Find the height of the ball at time $t$.

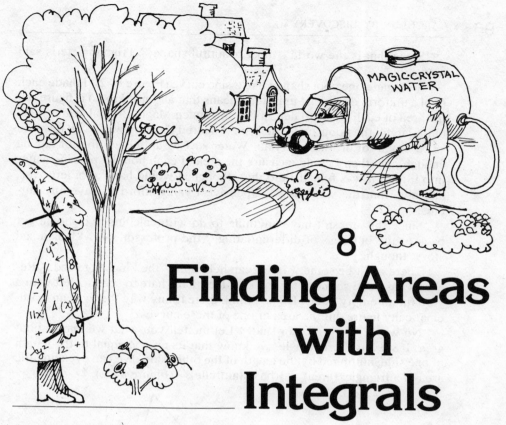

# 8
# Finding Areas
# with
# Integrals

Two days later, while we were meeting in the Main Conference Room, Recordis again had a problem.

"I can't take these terrible businessmen," he complained. "It's appalling. We've got to do something about this inflation."

"What are you talking about?" the professor asked.

"They raised the rates for Magic Crystal Water, and now I'm stuck. I can only afford the exact amount of Magic Crystal Water that I need."

"What do you need Magic Crystal Water for?" the king asked.

"Remember the house I built in the country that we had so much trouble designing?" We all nodded. "I decided to decorate the yard with some nice pools, made out of geometric figures. I want to fill the pools with Magic Crystal Water. I designed three curves with parabolic shapes (Figure 8–1), and Trigonometeris told me that I should make a pool shaped like a sine curve."

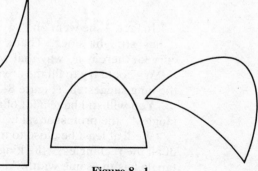

**Figure 8–1.**

95

"Since that is the world's most beautiful shape," Trigonometeris said modestly.

"So I made one pool shaped like a sine curve (Figure 8–2). I made each pool a uniform depth of 1 unit; that means that all I need to do is figure out the area of each curve to determine how much Magic Crystal Water is required to fill the pools up to the brim. I figured the only way to do it would be to buy a lot of Magic Crystal Water so I would be assured of having enough, and then I could measure the area of each pool by pouring water into it until it was full. If I did that, though, I would have a lot left over, and now that the water is so expensive I can't afford to have any surplus."

"Since this doesn't have anything to do with calculus, we should get back to the business of differentiating," the professor said. "It is a sad story, though."

"There must be some way we can help him," the king protested. "Recordis has done so much for the kingdom that I hate to see him left with a yard full of empty holes. Do you think there is any way that we can mathematically figure out the area of one of these curves?"

"No way!" the professor said. "Let me tell you what we know about area. If we have a rectangle, we know that its area is equal to the length of one side multiplied by the length of the other side. We can even find the area of a triangle: (area) = ½(base)(altitude)." (Figure 8–3.)

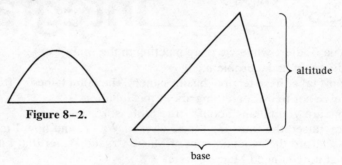

**Figure 8–2.**

altitude

base

**Figure 8–3.**

"In fact," she went on, "we can find the area of anything just so long as it has straight sides. That's all geometry. But if we have something curved, there is no way that we can find the area of it."

"We could try to fill the area of the curve with a lot of little rectangles," the king suggested. (Figure 8–4.)

"You will still have a lot of area left over that is not included in the rectangles!" the professor said.

"We'll at least be close to the area of the curve if we figure out the area of all the rectangles," the king pointed out. "Let's suppose that each rectangle has the same width."

"We can call that $\Delta x$," Recordis said.

"We need to figure out the boundaries of the region," the king noted.

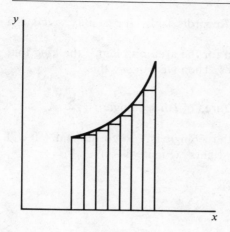

**Figure 8–4.**     **Figure 8–5.**

"Let's call the left-hand boundary $x = a$, the right-hand boundary $x = b$, and the lower boundary $y = 0$." (Figure 8–5.)

"We can tell what the area of the first rectangle is," Recordis said.

$$(\text{area})_1 = (\text{width}) \times (\text{height}) = f(a)\,\Delta x$$

We also found the areas of the second and third rectangles:

$$(\text{area})_2 = f(a + \Delta x)\,\Delta x$$

$$(\text{area})_3 = f(a + 2\Delta x)\,\Delta x$$

"And the total area under the curve is the sum of the first area plus the second area plus the third area plus . . . ," the king said.

"Hold it!" Recordis protested. "You must have a hundred rectangles up there, and if you want me to add up the areas of all of them you will have to pay me about a hundred times what you're paying me now!"

"Don't we have a shorter way for writing a sum like that?" Trigonometeris asked.

"Of course!" the king said. "Summation notation! Remember that we used a crooked letter $s$, $\Sigma$. (The symbol $\Sigma$ is the Greek capital letter sigma.) We put where to start at the bottom: $\sum_{i=1}$, and where to stop at the top: $\sum^{10}$, and we put what we want to add up along the sides: $\sum_{i=1}^{10} i^2$. For example, we might have the following."

$$\sum_{i=1}^{10} i = 1 + 2 + 3 + 4 + 5 + 6 + 7 + 8 + 9 + 10 = 55$$

$$\sum_{i=1}^{10} 2i = 2 + 4 + 6 + 8 + 10 + 12 + 14 + 16 + 18 + 20 = 110$$

$$\sum_{i=1}^{10} i^2 = 1 + 4 + 9 + 16 + 25 + 36 + 49 + 64 + 81 + 100 = 385$$

"I remember how that worked," Recordis said. "It certainly saved us a lot of writing."

"We can use summation notation for the area problem," the king told him. "Suppose we have $n$ rectangles. Then we can say this."

$$\text{(area of all rectangles)} = \sum_{i=1}^{n} \text{(area of } i\text{th rectangle)} = \sum_{i=1}^{n} A_i$$

"We know what the area of the $i$th rectangle is," Recordis said. "It will be equal to $f(x_i)\, \Delta x$, if we define $x_i$ right." (Figure 8–6.)

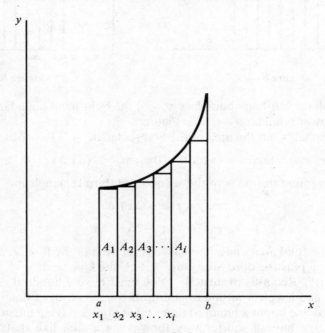

**Figure 8–6.**

Igor wrote the equation for the area of all the rectangles:

$$\text{(area of rectangles)} = \sum_{i=1}^{n} A_i = \sum_{i=1}^{n} f(x_i)\, \Delta x$$

"We still have a problem," Recordis said. "The summation notation makes it simple to write a sum involving hundreds of terms, but it doesn't make it any easier to do the actual hard work of adding them all together. And that's what I always get stuck doing."

"We have an even bigger problem," the professor pointed out. "This expression gives us the area of the rectangles, but it still doesn't give us the area of the curve."

"I know how we could get closer to the curve's area," the king said. "We could use twice as many rectangles. Then there would be much less difference between the area under the curve and the area of all the rectangles.''

"You could get closer, but you still couldn't do it!" the professor objected. "There still is some area wasted between the curve and the rectangles. You'd have to use an almost infinite number of rectangles before you could get the right area."

"That's it!" the king shouted. "We'll use an almost infinite number of rectangles! We'll say that the area of the curve is the *limit* of the sum of the areas of all the rectangles as we let the number of rectangles go to infinity."

$$\text{(area under curve)} = \lim_{\Delta x \to 0, \, n \to \infty} \sum_{i=1}^{n} f(x_i) \, \Delta x$$

"You can't take a limit to infinity!" the professor said.

"I just did," the king responded.

"But look what will happen!" the professor went on. "Suppose $n = \infty$ and $\Delta x = 0$. Then $f(x_i) \, \Delta x$ would be zero all the time. You would be adding together an infinite number of zeros."

"But we can't ever let $\Delta x$ actually equal zero," the king said. "We can let it become absolutely just-about-there close to zero. That's what 'limit' means."

"We used limits when we found derivatives," Recordis added. "We let $\Delta x$ come very close to zero, but we never let it actually equal zero."

We were all very impressed until the professor suddenly realized that we had not come much closer to solving the problem. "This still does not tell us how to take a given curve, $y = f(x)$, and two given numbers, $a$ and $b$, and come up with a number that is equal to the area under the curve."

We realized that our definition of the derivative in terms of limits was simple enough, because we could easily calculate the actual numerical value of the limit for a specific function. The area limit involved a sum with an infinite number of terms, so we all realized there was no way that we could directly figure out the area by adding all these terms together. "We couldn't solve this problem if we stayed here until Hotspot Caves freeze over," the king said sadly.

"We'll have to give up," the professor mourned.

"I'll find something else to do with my pools," Recordis said.

We were about ready to turn to the business of differentiation and integration when there came a sudden swoosh! through the window. There was an evil, cackling laugh, and the next thing we saw was . . . the gremlin!

"This time I have you by the throat," he laughed. "So you are about to give up. I suggest that you do. Surrender to my supreme powers of evil." He held out his cape, and we could see misty images of Recordis' yard and the pools with the algebraic shapes. As we watched, a fire kindled in each pool.

"These fires are designed so the flames can sweep across the entire kingdom of Carmorra,'' the gremlin went on. ''There is only one thing that can put these fires out—Magic Crystal Water. But you must use *exactly* the right amount. One drop too little in any pool—and the flames will still escape. One drop too much—and a giant, sizzling, steaming flood will completely engulf Carmorra! I suggest that you give up now, and submit yourselves to my becoming King of Carmorra!''

''Never!'' the king cried.

''We certainly cannot allow that!'' Recordis said.

''Unless you pour in exactly the right amount of Magic Crystal Water the flames will escape at exactly sunset,'' the gremlin laughed wickedly. He looked at his wrist hourglass. ''You have exactly 2 hours and 56 minutes.'' Before we knew what was happening, he had whipped his cape around himself and, with a tremendous blast of hot air, had disappeared out the window again.

''I think we had better not give up,'' the professor said.

''I wish this was a calculus problem!'' Recordis moaned. ''We know how to solve those.''

We spent a long time trying to calculate the limits directly, but we made no progress. We tried desperately to come up with another method that might work. Finally the king decided to put Builder to work. Builder set up a station at Recordis' house, and the king ordered a huge barrel of Magic Crystal Water to be paid for out of the nation's treasury since, after all, this problem now involved the defense of Carmorra. We worked out a signaling system so that, as soon as we came up with the answer for the area of the pools, Builder would know right away and could pour exactly the right amount of water into the pools. Still, another half-hour of work on the area problem failed to produce any results.

''We could give the kingdom up to the gremlin,'' Recordis said.

''That would spoil everything!'' the king objected.

''But we can't let ourselves be burned to death,'' the professor said.

''Or flooded to death,'' Trigonometeris added.

I tried a couple of ideas that didn't work. Finally I set out on a desperation path. Igor was doing algebra as fast as any Visiomatic Picture Chalkboard Machine could, but our time was quickly slipping way.

''Let's assume that there exists some function—call it $A(x)$—that will tell us the area under the curve between the lines $x = a$ and $x = x$,'' I said. (Figure 8–7.)

''How do we know that there is a function like that?'' Recordis protested.

''We have to assume that there is one,'' I said. ''If there isn't, then we're lost already. If there is, maybe we can see how it behaves.''

''We can tell what $A(a)$ is,'' the professor noted helpfully. ''$A(a)$ must be zero, because there is no area between a line and itself.''

''I know what $A(b)$ is,'' Recordis said. ''That's the answer we're looking for—the area under the entire curve.''

''That's obvious,'' the professor sniffed.

"We have only 9 minutes left," Recordis said. "Now we're doomed."

"Let's figure out how the function changes at different points," I said, as a sudden inspiration struck me. "Let's find out what the area function is equal to at another point, say $x + \Delta x$ (Figure 8–8). This solid area is what we know to be $A(x)$. This striped rectangle, when added to the solid area, gives $A(x + \Delta x)$."

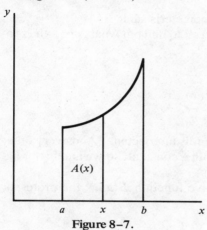

Figure 8–7.

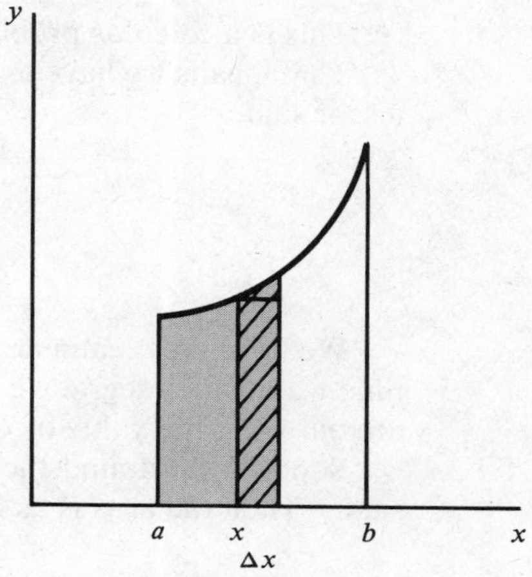

Figure 8–8.

"We can figure out what the solid area is," the king pointed out. "It looks like one of the rectangles we were using earlier."

$$(\text{striped area}) \simeq (\text{width}) \times (\text{height})$$

$$\simeq f(x)\,\Delta x$$

$$A(x + \Delta x) = (\text{striped area}) + (\text{solid area})$$

$$\simeq f(x)\,\Delta x + A(x)$$

"We also know that the smaller $\Delta x$ becomes, the closer the area of the rectangle approaches the striped area," the professor said.

"Let's put in the limit as $\Delta x$ goes to zero," the king suggested. "Then we will have the exact area."

$$\lim_{\Delta x \to 0} A(x + \Delta x) - A(x) = \lim_{\Delta x \to 0} f(x)\,\Delta x$$

"We have 6 minutes left," Recordis said. "Now we're *really* doomed."

All of a sudden, we each struck upon the same idea.

"Do you see what I see?" the professor asked.

Very slowly the king suggested, "Let's take each side and divide it by $\Delta x$."

$$\lim_{\Delta x \to 0} \frac{A(x + \Delta x) - A(x)}{\Delta x} = \lim_{\Delta x \to 0} f(x)$$

$$= f(x)$$

"We know what the left-hand side of that equation is," the professor said. "It fits right in with our definition of the derivative, so that means that the left-hand side of the equation is the derivative of the area function with respect to $x$."

$$\frac{dA}{dx} = f(x)$$

"This is a calculus problem!" Trigonometeris said.

"That means we have to take an integral to find out what $A$ is," the professor said.

$$dA = f(x) \; dx$$

$$\int dA = \int f(x) \; dx$$

$$A = \int f(x) \; dx$$

"We have only 3 minutes left!" Recordis interjected. "And every other time we did an integral we ended up with a constant, so we had better figure out what the value of $C$ will be."

"Suppose we found the antiderivative function $F(x)$," the professor said. "Then the area is as follows."

$$A(x) = F(x) + C$$

"We need to find an initial condition to tell us what $C$ is," the king remarked.

"I know an initial condition," Recordis said. "We said that $A(a) = 0$."

$$A(a) = 0 = F(a) + C$$

$$C = -F(a)$$

"So that means $A(x) = F(x) - F(a)$," the professor contributed.

"And therefore," the king said, "we have this."

$$A(b) = \text{(total area under curve)} = F(b) - F(a)$$

As fast as he could write, Recordis jotted down what he called the *fundamental theorem of integral calculus* (see page 103).

"Let's think of an easy way to write the area in terms of an integral," the professor said quickly. "Let's write the integral, and then write the boundary terms like this."

$$A = \int f(x) \; dx \; \Big|_a^b = F(b) - F(a)$$

(The boundary terms $a$ and $b$ are usually called the *limits of integration.*)

"Let's forget that little vertical line and write it like this," Recordis said.

$$A = \int_a^b f(x) \; dx = F(b) - F(a)$$

## FUNDAMENTAL THEOREM OF INTEGRAL CALCULUS

The area below the curve $y = f(x)$, above the line $y = 0$, to the right of the line $x = a$, and to the left of the line $x = b$ equals $A = F(b) - F(a)$, where $F(x)$ is the antiderivative function such that $dF(x)/dx = f(x)$. (See Figure 8–9.)

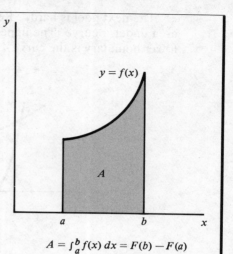

**Figure 8–9.**      $A = \int_a^b f(x)\, dx = F(b) - F(a)$

"That integral will tell us exactly what number is equal to the area."

"That is a definite integral if I ever saw one," Trigonometeris observed.

"All right, we'll call it a *definite integral* if it stands for an area," the professor said. "A definite integral will look almost the same as an indefinite integral, except that we have two integration limits written next to the integral sign. Now we've got to solve these problems. Recordis, what are the equations of your curves?"

"The first one is simple," Recordis said. "The area is the area under the function $y = x^2 + 5$ from $x = 3$ to $x = 6$."

We set up the definite integral, with the limits of integration written right next to the integral sign.

$$A = \int_3^6 (x^2 + 5)\, dx$$
$$= \tfrac{1}{3} x^3 \Big|_3^6 + 5x \Big|_3^6$$
$$= (\tfrac{1}{3})\, 6^3 - (\tfrac{1}{3})\, 3^3 + 5 \cdot 6 - 5 \cdot 3$$
$$= \tfrac{216}{3} - \tfrac{27}{3} + 30 - 15$$
$$A = 78$$

"The next one is interesting," Recordis said. "The curve is $f(x) = -x^2 + 9$, with the boundary points $-3$ and $3$."

$$A = \int_{-3}^{3} (-x^2 + 9)\, dx$$
$$= -\tfrac{1}{3} x^3 \Big|_{-3}^{3} + 9x \Big|_{-3}^{3}$$
$$= -\tfrac{27}{3} - \tfrac{27}{3} + 27 + 27 = 54 - 18 = 36$$

"The next pool is hard," Recordis told us. "Its area is not equal to the area under a curve. The upper boundary is the curve $x^2/25 + 2$, and the lower boundary is the curve $4x^2/25 - 1$." (Figure 8–10.)

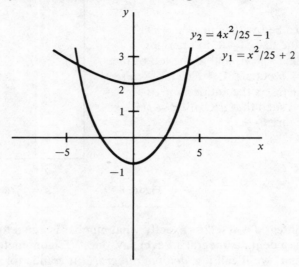

**Figure 8–10.**

"We can find the area between two curves," the professor said quickly. "All we need to do is define a new function equal to the *difference* between the two curves."

$$y_1 = \frac{x^2}{25} + 2$$

$$y_2 = \frac{4x^2}{25} - 1$$

"Let $y = y_1 - y_2 = x^2/25 + 2 - (4x^2/25 - 1)$. Now we can integrate that function from $-5$ to $5$."

$$A = \int_{-5}^{5} (y_1 - y_2)\, dx = \int_{-5}^{5} \left( \frac{-3x^2}{25} + 3 \right) dx$$

$$= -\frac{1}{25} x^3 \Big|_{-5}^{5} + 3x \Big|_{-5}^{5}$$

$$= -5 - 5 + 15 + 15 = 30 - 10 = 20$$

"The area of the first curve is 78," the professor cried over the signaling system. "The other two curves have areas of 36 and 20." Builder quickly signaled back that he had poured the exact amount of Magic Crystal Water into those three pools, and that the fires were out.

"We have only 1 minute left!" Recordis said. "There still are flames in the pool shaped like a sine curve!" The sun had almost completely disappeared behind the mountains of western Carmorra.

"We need the integral of $y = \sin x$, from $x = 0$ to $x = \pi$," the professor said.

$$A = \int_0^\pi \sin x \, dx = -\cos x \, \Big|_0^\pi = -\cos \pi + \cos 0$$

$$= -(-1) + 1 = 2$$

"Ten seconds left!" Recordis shouted.

"The area's 2! Exactly 2 square units!" the professor signaled Builder.

We all closed our eyes. The last shadow disappeared as the mountains blocked out the sun. Limp with fear, we waited for the flames to engulf us, but Builder had done his job in time.

We were safe!

_____ Note to Chapter 8

Notice that the definite integral $\int_a^b f(x) \, dx$ represents the area under the curve $f(x)$ only if $b > a$ and $f(x) > 0$ for all $a < x < b$. If the function is negative everywhere in the interval from $x = a$ to $x = b$, then the value of the definite integral will be the negative of the area between the curve and the $x$ axis. If the function is positive at some places and negative at other places in the interval $x = a$ to $x = b$, then the value of the definite integral will be equal to the total area under the positive part of the curve minus the total area between the negative part of the curve and the $x$ axis. The final result may be positive, negative, or zero. Sometimes you will be interested in the algebraic value of the definite integral, in which case you really will want the negative areas to cancel out the positive areas.

_____ Exercises

1. Find the area under the curve $y = \cos x$ from $x = -\pi/2$ to $x = \pi/2$.
2. Find the area under the curve $y = \sin^2 x$ from $x = 0$ to $x = \pi$.
3. Find the area between the $x$ axis and one arch of the curve $y = 3 \sin 5x$.
4. Find the area between the parabola $y = 2x^2 - 8$, the line $x = 2$, the line $x = -2$, and the $x$ axis.

Evaluate the following. (Remember that the answer to a definite integral is a number.)

5. $\int_0^1 (x^2 + 3x + 5) \, dx$

6. $\int_{-3}^5 (9x + 10) \, dx$

7. $\int_{d_1}^{d_2} (ax^2 + bx + c) \, dx$

8. $\displaystyle\int_{d_1}^{d_2} (at^2 + bt + c)\, dt$

9. $\displaystyle\int_0^a 14\, dx$

10. $\displaystyle\int_{-1}^1 (x^2 + 1)\, dx$

11. $\displaystyle\int_0^1 (5x^5 + 10x^3 + 3x)\, dx$

12. $\displaystyle\int_0^1 (4x^3 + 3x^2 + 2x + 1)\, dx$

13. $\displaystyle\int_0^a (x^3/3 + x^2/2 + x)\, dx$

14. $\displaystyle\int_{d_1}^{d_2} (x^m + x^n)\, dx$

15. $\displaystyle\int_0^1 [(m + 1)x^m + (n + 1)x^n + (p + 1)x^p]\, dx$

16. $\displaystyle\int_{-1}^1 x^{100}\, dx$

17. $\displaystyle\int_0^{\pi/4} \sin\theta\, d\theta$

18. $\displaystyle\int_{-\pi/4}^{\pi/4} \cos\theta\, d\theta$

19. $\displaystyle\int_1^2 (x^{-4} + x^{-3})\, dx$

20. $\displaystyle\int_1^2 (x^2 + x^{-2})\, dx$

21. Use definite integrals to show that the area of a triangle is given by $\frac{1}{2}$(base)(altitude).

22. Show that $\displaystyle\int_a^b f(x)\, dx = -\int_b^a f(x)\, dx$.

23. Evaluate the definite integral $y = \displaystyle\int_0^{2\pi} \sin\theta\, d\theta$.

24. Use the table of values (Table 8–1) to estimate the area under the curve $y = \sin x$ from $x = 0$ to $x = \pi/2$, using (a) two inscribed rectangles and (b) eight inscribed rectangles.

**Table 8–1**

| $x$ | $\sin x$ |
| --- | --- |
| 0.1745 | 0.1736 |
| 0.3491 | 0.3421 |
| 0.5236 | 0.5000 |
| 0.6981 | 0.6428 |
| 0.8727 | 0.7661 |
| 1.0472 | 0.8660 |
| 1.2217 | 0.9397 |
| 1.3963 | 0.9848 |
| 1.5708 | 1.0000 |

25. The definite integral can be used to find the average value of a function over a given integral. The average value of the function $y = f(x)$ over the interval $x = a$ to $x = b$ is defined to be $y = [1/(b - a)]$ $\int_a^b f(x)\ dx$. Find the average value of the function $y = x^2$ in the interval $x = 1$ to $x = 3$.

26. Find the average value of the function $y = x^{-2}$ over the interval $x = 1$ to $x = 4$.

27. The voltage in an alternating-current circuit is described by the function $V(t) = A \sin \omega t$. The peak-to-peak voltage ($V_{pp}$) is defined to be $V_{pp} = 2A$. The rms (for root-mean-square) voltage is a measure of the average value of the voltage in the circuit. It is defined to be $V_{rms} = \sqrt{\{|V(t)|^2\}_{average}}$. Find the average value of $V^2$ over the interval $t = 0$ to $t = \pi/\omega$. Then take the square root of the result to find $V_{rms}$. (Remember that $\sin^2 \theta = \frac{1}{2}(1 - \cos 2\theta)$.) What is $V_{pp}$ when $V_{rms} = 1$?

28. The area under the parabola $y = x^2$ from $x = 0$ to $x = a$ is given by the expression:

$$A = \lim_{n \to \infty} \sum_{i=1}^{n} (x_i)^2\ \Delta x$$

You can evaluate this summation directly, if you know this tricky formula:

$$\sum_{i=1}^{n} i^2 = \frac{n}{6}\ (n + 1)(2n + 1)$$

Evaluate the summation, and compare the result with the definite integral $\int_0^a x^2\ dx$.

# 9
# Natural Logarithms

The gremlin disappeared after his defeat, and we heard no more from him for some time. The whole kingdom was excited by our discovery of the method for finding areas.

"We should expand our business," the professor said. "We should call it the Differentiation and Integration Business."

"Now we can do just about anything!" Recordis boasted. "We can find areas or velocities-given-positions or positions-given-velocities or the slopes of tangents."

Business was great for the next few days. The first time we hit a snag occurred one morning when I was at the Differentiation and Integration Business office with Recordis. We were approached by a tall, elegantly dressed customer. "That's my neighbor, Count Q," Recordis whispered in my ear. "After we solve his problem, he will probably give us a gift large enough to support our business for a year."

"What can I do for you?" Recordis asked, making a special effort to impress the count.

"I have a horrible problem," the count said. "My daughter decided that she wanted to construct a life-size model of Hasselbluff Mountain, entirely out of beads. I was able to convince her to accept a one-tenth scale model, but that still required a huge amount of beads—4 million grams, as it turned out."

"Four million grams of beads!" Recordis breathed in astonishment.

"I had a large bead container built to hold them, but, as you might expect, my daughter knocked the container over this morning. The beads are all over the yard now. Fortunately, Rutherford, your dog, decided that it would be fun to pick them up. He started this morning. He does have a lot of energy, and he has been working fairly fast."

"What's the problem?" Recordis asked.

"I need to know how long it will take him to pick up the beads. At first the beads were easy to pick up, and he was able to scoop lots of them into the container every minute. As there are fewer and fewer beads, though, they're harder to pick up, so he can't pick them up as fast. I measured the rate at which he can pick up the beads, and I found out that each hour he can pick up one quarter of the beads that are in the yard at the start of the hour. Do you think you can find out how long it will take until he's finished?"

"Certainly," Recordis said proudly. "We invented an integral, symbolized by a squiggle: $\int$. Suppose you tell me what $dx/dt$ is (that means the rate at which some variable $x$ is changing with respect to some variable $t$). In your case $x$ would be the number of beads left in the yard at a given time.

"If you tell us $dx/dt = f(t)$, where $f(\ )$ is some function of time, then we can easily calculate $x$ as a function of $t$."

$$dx = f(t)\ dt$$

$$\int dx = \int f(t)\ dt$$

$$x = \int f(t)\ dt$$

Recordis was enjoying being able to show off his knowledge, but I could see that Count Q was becoming impatient.

"We can also do it this way, if you tell us $dx/dt = f(x)$."

$$\frac{dx}{dt} = f(x)$$

$$\frac{1}{f(x)}\ dx = dt$$

$$t = \int \frac{1}{f(x)}\ dx$$

"In your case we have $dx/dt = -\frac{1}{4}x$. So let's do this," Recordis said with a flourish.

$$-4\ \frac{dx}{dt} = x$$

$$-4\ \frac{1}{x}\ dx = dt$$

$$\int -4\ \frac{1}{x}\ dx = \int dt$$

"Then we use our Perfect Integral Rule," Recordis said, making sure that the count could hear the capital letters he had just given the name.

$$t = \int -4 \frac{1}{x}\, dx$$

"And our Multiplication Rule," he continued.

$$t = -4 \int x^{-1}\, dx$$

"Now we have our Power Rule."

$$t = -4 \frac{1}{(-1) + 1} x^{-1+1} + C$$

As Recordis was about to put the grand finishing touch on the problem, he suddenly choked and a cold, desperate sweat broke out over his face. "Ah, yes, as I was saying, this is a very simple insert-the-numbers problem." He suddenly turned to me. "Why don't you tell the count why we'll have the answer for him tomorrow rather than today?"

I suddenly saw what Recordis had seen. "We have to go to lunch right now," I said.

"Lunch right now?" the count cried. "It's 11:23½ A.M.!"

"We always have lunch at 11:23½ A.M., on the dot," Recordis said. He quickly pulled the curtain in front of the Differentiation and Integration Business and ran into the palace.

"Help! Help! Meet right away!" Recordis shouted. A few minutes later we were all gathered in the Main Conference Room. Recordis told everyone our problem.

"The power rule doesn't work if $n = -1$!" he cried. "I have never been so embarrassed in my life."

"Calm down," the professor said. "I'm sure we can figure something out. Igor, write down the power rule."

$$\int x^n\, dx = \frac{1}{n+1} x^{n+1} + C$$

"Just try putting $n = -1$ in there!" Recordis yelled.

"You get 1/0 times $x^0$, which is 1/0 times 1, or 1/0," the professor said.

"But that doesn't mean anything!" the king protested. "We found that out in algebra."

"Also we can see that it doesn't work," Trigonometeris said. "It is obvious that $(d/dx)(1/0)$ does not equal $1/x$."

"We better add that condition to the power rule," the king said.

---

**POWER RULE**

$$\int x^n\, dx = \frac{1}{n+1} x^{n+1} + C \qquad \text{if } n \neq -1$$

"That must mean that there is no such thing as a function with derivative $x^{-1}$," the professor said. "If there is no such function, then the integral $\int x^{-1}\,dx$ does not exist. I can't think of any reason why we should worry about that integral."

"But there is such a function!" Recordis told him. He explained the problem with the beads. He also said that if they could find an answer by that night they would probably receive a generous gift from the count.

"Maybe we can use a definite integral to help us out," the king said. "We found that an integral represents the area under the curve, so let's draw a curve and find the area under it."

Igor drew a graph of the function $f(q) = 1/q$. (See Figure 9–1.)

"I know what this definite integral means," the professor said.

$$\int_a^b f(q)\,dq \; = \; \int_a^b \frac{1}{q}\,dq \; = \; \text{area above the } q \text{ axis,}$$
$$\text{below the curve } f(q) = 1/q,$$
$$\text{to the left of the line } q = b,$$
$$\text{and to the right of the line } q = a$$

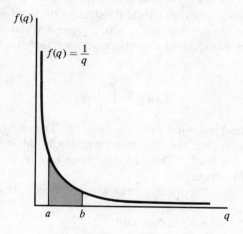

**Figure 9–1.**

"The integral function must exist!" Recordis said. "That area is clearly some real number that you could measure if you had to."

"All we have to do is set up a Mysterious Function that gives us this area and investigate what its properties are," the professor said. "If we find a function that gives us the area, then that means we have also found the function that gives us the antiderivative we need."

"That's from the fundamental theorem of integral calculus," Recordis added.

"We should fix the left-hand boundary of the area," the king said. "Then the area will be a function of just one variable: the right-hand boundary."

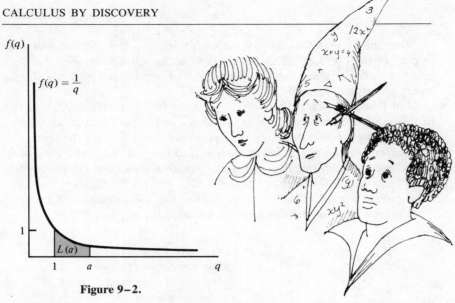

**Figure 9-2.**

We decided to fix the left-hand boundary at $q = 1$, since it was clear that the Mysterious Function would do strange things if we fixed the left-hand boundary at the point where $q = 0$. I suggested that we use the letter $L$ to stand for the Mysterious Function, so we made the following definition (Figure 9-2):

$$L(a) = \int_1^a \frac{1}{q} \, dq$$

"How do we find out what this function is?" Trigonometeris asked.

"I would suggest that the first thing we should do is look for properties," the professor said. "Does anybody see any obvious properties of the Mysterious Function?"

"I see one obvious property," the king replied. "It looks as though $L(1) = 0$."

"That's a start," the professor said.

"We can also say that, when $a$ is greater than 1, $L(a)$ is greater than 0," Recordis said.

"And from the way we set up our convention for the sign of an area, we can say that, when $a$ is less than 1, $L(a)$ is less than 0," Trigonometeris added. (See the note to Chapter 8, and exercise 8-22.)

"What happens if $a$ is less than or equal to 0?" Recordis asked.

"Then everything would blow up!" the professor said. "You couldn't define the area, so we may just as well say that $L(a)$ is undefined when $a$ is less than 0."

We made a list of the properties we had found (see page 113).

"Does anybody recognize these properties?" the professor asked.

"I do!" the king said. "They do look familiar! The answer's right at the tip of my tongue, but I can't think of it."

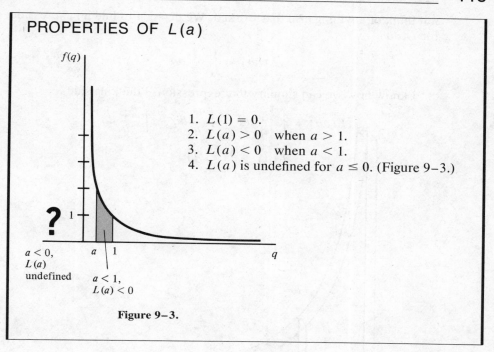

PROPERTIES OF $L(a)$

1. $L(1) = 0$.
2. $L(a) > 0$   when $a > 1$.
3. $L(a) < 0$   when $a < 1$.
4. $L(a)$ is undefined for $a \le 0$. (Figure 9–3.)

$a < 0$,
$L(a)$
undefined

$a < 1$,
$L(a) < 0$

**Figure 9–3.**

Nobody else could think of any algebraic or trigonometric functions that fit these properties, but Trigonometeris came up with an idea that would allow us to calculate values for $L(a)$.

"Remember when we invented my functions—the trigonometric functions? We decided that they were more complicated than ordinary functions, so we constructed my triangles so we could measure the sine, cosine, or tangent of any angle. After we made the triangles, we drew up a table of values so we could just look in the table for the trigonometric functions of any given angle. All we have to do now is build something to calculate $L(a)$, and then make a table of values. Let's have Builder build a pool with a sliding wall (Figure 9–4). All he has to do is set the sliding wall at different values of $a$, and then measure the amount of water in the pool."

Builder was immediately summoned to the conference room. He was told the urgency of the project, and he promised to put his best craftsmen to work right away.

"Make sure the pool is accurate!" the professor called after him as he went away.

We ate lunch while we waited for Builder to bring us a table of values. After a while we decided to look for some more properties. We tried to find an expression for $L(a + b)$, where $a$ and $b$ are any two numbers, but we soon gave up. We quickly established that $L(a + b)$ does not equal $L(a) + L(b)$ (let $a = b = 1$, and see what happens), but we could not think of anything else that worked. We decided to find an expression for $L(ab)$.

$$L(ab) = \int_1^{ab} \frac{1}{q} \, dq$$

"I know how we can simplify that expression," the king said.

$$L(ab) = \int_1^{ab} \frac{1}{q} \, dq = \int_1^{a} \frac{1}{q} \, dq + \int_a^{ab} \frac{1}{q} \, dq$$

$$= L(a) + \int_a^{ab} \frac{1}{q} \, dq$$

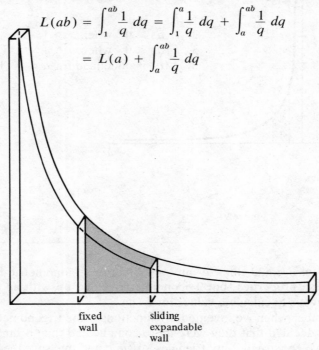

fixed
wall

sliding
expandable
wall

**Figure 9–4.**

"How does that help?" the professor said. "We don't know what that second term is."

"If we could change the $a$ in the lower bound into a 1, we would be in better shape," Recordis remarked.

"Maybe we can do that," the king said. "We can make a substitution, so that instead of $q$ as the integration variable we have some other variable."

Recordis suggested that we try $u = q/a$, $q = au$.

$$L(ab) = L(a) + \int_{q=a}^{q=ab} \frac{1}{au} \, dq$$

"Now we can't solve it because the integrand is in terms of $u$ while the boundary limits and the $d$-variable terms are still in terms of $q$."

"Then we will have to change them so they are in terms of $u$," the king said. "Since $q = au$, we know that $dq = a \, du$."

"And I can't see any reason why we can't change the limits so that they are in terms of $u$," Recordis added. "If $u = q/a$, then $u = b$ when $q = ab$ and $u = 1$ when $q = a$."

We put these values in the integral:

$$L(b) = L(a) + \int_{q=a}^{q=ab} \frac{1}{au}\, dq = L(a) + \int_{u=1}^{u=b} \frac{1}{au}\, a\, du$$

"Are you sure that is the same integral we started with?" the professor asked suspiciously.

"It should be," the king said. "All we've done is change the name of a variable."

We made a list of the procedure for the substitution method for definite integrals.

---

## SUBSTITUTION METHOD FOR DEFINITE INTEGRALS

Some integrals can be made simpler by substituting some variable $u$ in place of the original variable of integration. The change must be made in three places:

1. The middle of the integral (the integrand) must be written in terms of the new variable $u$.
2. The $d$-variable (or differential) term must be written in terms of the new variable ($du$).
3. The two limits of integration must be changed from limiting values of $q$ to limiting values of $u$.

---

"Now where were we?" Recordis said.

$$L(ab) = L(a) + \int_1^b \frac{a\, du}{au}$$

"We can cancel out the $a$'s," the professor suggested.

$$L(ab) = L(a) + \int_1^b \frac{du}{u}$$

"We know what the second term is!" she said suddenly. "That's the definition of $L(b)$!"

$$L(ab) = L(a) + L(b)$$

"That's an amazingly simple property for this function," the professor added.

"Didn't we see this property before?" Recordis said, leafing through his old algebra book.

"I recognize it!" the king exclaimed triumphantly. "Remember the logarithm function!"

"The logarithm function!" the professor said in awe.

"The logarithm function!" Trigonometeris said in awe.

"The logarithm function!" Recordis said in awe. "What's the logarithm function?"

"Don't you remember?" the professor chided. "We said that, if $y^x = a$, then $x = \log_y a$ (this means that $x$ is the logarithm to the base $y$ of $a$)." The professor pointed to a page in the book that described the logarithm function.

"Oh," Recordis said. "Now I remember." (The people in Carmorra had developed a fairly extensive list of properties of the logarithm function. If you are not familiar with logarithms to the base 10, you can consult a book on algebra.)

"And the logarithm function has all the properties we found for our mysterious function," the king pointed out.

$$\log 1 = 0$$

$$\log a > 0 \qquad \text{when } a > 1$$

$$\log a < 0 \qquad \text{when } a < 1$$

$$\log a \text{ is not defined for } a \leq 0$$

$$\log(ab) = \log a + \log b$$

"But there is something else about the logarithm function," Recordis said. "We have to specify a *base*. We decided that, if we wrote log $a$ without explicitly stating a base, we meant the logarithm to the base 10 ($\log a = \log_{10} a$), and we called that the *common logarithm*. But we could also take logarithms to any base: $\log_2 a$, $\log_{100} a$, or whatever."

"Except for the number 1," the professor added.

"That's right," Recordis said. "We found that there is no such thing as a logarithm to the base 1 ($\log_1 a$). But that doesn't help us know what base to use for the Mysterious Function, $L(a)$."

"We'll have to wait until Builder has brought us a table of values," the king stated. "Then we should be able to tell what base it is."

"I hope it's some easy number like 2 or 3," Recordis said. "We need to think of some letter to stand for the unknown base until we find out what it is."

They turned to me, and I came up with another suggestion: that we let the letter $e$ stand for the unknown base.

"We still have to check this," the professor said. "You are saying that *integrating* a $1/x$ function gives you a logarithm function. If that's true, then *differentiating* a logarithm function will give you a $1/x$ function."

We set up the definition of the derivative:

$$y = \log_e x$$

$$y + \Delta y = \log_e(x + \Delta x)$$

$$\Delta y = \log_e(x + \Delta x) - \log_e x$$

"We should have done this a long time ago," Recordis remarked. "I thought that we had found the derivative of every single possible function already, but I completely forgot about the logarithm function."

"We can use a logarithm property here," the king said. "Remember that log $a$ − log $b$ = log($a/b$)."

$$\Delta y = \log_e \left( \frac{x + \Delta x}{x} \right)$$

"We had better divide both sides by $\Delta x$," the professor advised. "That seems to be the standard way to proceed under these circumstances."

$$\frac{\Delta y}{\Delta x} = \frac{\log_e(1 + \Delta x/x)}{\Delta x}$$

"Now we're stuck," Recordis said.

We puzzled over this situation for several minutes, until I had a sudden inspiration. "Since we would like to end up with $1/x$, why don't we try multiplying the numerator and the denominator by $x$? That way we'll have at least one $x$ in the denominator."

"It can't hurt too much," Recordis said.

$$\frac{\Delta y}{\Delta x} = \frac{x \log_e (1 + \Delta x/x)}{x \, \Delta x}$$

$$= \frac{1}{x} \frac{x}{\Delta x} \log_e \left( 1 + \frac{\Delta x}{x} \right)$$

"Somewhere along in here you will need to put a limit as $\Delta x$ goes to zero on both sides," Recordis pointed out.

$$\lim_{\Delta x \to 0} \frac{\Delta y}{\Delta x} = \frac{dy}{dx} = \lim_{\Delta x \to 0} \frac{1}{x} \frac{x}{\Delta x} \log_e \left( 1 + \frac{\Delta x}{x} \right)$$

Recordis wanted a shorter way to write this expression, so we made the definition $w = \Delta x/x$. Then we could rewrite the derivative:

$$\frac{dy}{dx} = \frac{1}{x} \lim_{w \to 0} \frac{1}{w} \log_e(1 + w)$$

"I remember another property of logarithms," the professor said. "We discovered that $n \log x = \log(x^n)$."

$$\frac{dy}{dx} = \frac{1}{x} \lim_{w \to 0} \log_e (1 + w)^{1/w}$$

$$= \frac{1}{x} \log_e \lim_{w \to 0} (1 + w)^{1/w}$$

"I have an idea," Recordis stated. "If we could make this weird term, $\log_e \lim_{w \to 0} (1 + w)^{1/w}$, equal 1, then we would have the answer we want: $dy/dx = 1/x$."

"That's obvious," the professor said. "We can't just make it be equal to 1, though."

"He does have something!" the king exclaimed. "We might be able to use this expression to track down the mysterious number $e$! Remember that $\log_{(\text{some base})}(\text{some base}) = 1$, always. Remember that $\log_2 2 = 1$, $\log_{10} 10 = 1$, and $\log_{17} 17 = 1$, because $2^1 = 2$, $10^1 = 10$, and $17^1 = 17$. So we should have $\log_e e = 1$! And we also want to have $\log_e \lim_{w \to 0} (1 + w)^{1/w} = 1$! That means $\lim_{w \to 0} (1 + w)^{1/w}$ equals $e$!"

"Of course!" Recordis said. "The old mysterious-number-$e$-caught-up-in-the-screwy-limit trick!"

"Can we calculate that?" the professor asked.

"It's just a matter of arithmetic," the king said. "We can't calculate $e$ if $w = 0$, because then we would get $e = 1^\infty$, which doesn't help much. But we can calculate $e$ for any value of $w$ close to zero."

Igor and Recordis soon came up with a table of values (Table 9–1).

"Just exactly what I was afraid of!" Recordis moaned. "Look, $e$ turns out to be some number between 2.7181 and 2.7196."

"But *why?*" the king asked. "Is there something special about a number that is approximately 2.718? I can't think of any reason why there should be."

"It must be an irrational number," Trigonometeris remarked. "I don't think you could ever find an exact decimal representation for it. Just like most trigonometric functions."

"Is there anything else like it?" the king asked in great distress. He believed very strongly that he should be an impartial ruler, and it bothered him that one number, more than any other, should be singled out as the base for the $L(x)$ function.

"I can think of only one other case," the professor said solemnly. "When we found the circumference of a circle, we decided that the circumference was $2\pi r$, where $r$ is the radius. We found there was a special number, which we called pi, because it seemed to be so fundamental that it worked for any circle."

**Table 9–1**

| $w$ | $(1 + w)^{1/w}$ |
| --- | --- |
| 0.5 | 2.25 |
| 0.1 | 2.5937 |
| 0.01 | 2.7048 |
| 0.0001 | 2.7181 |
| −0.001 | 2.7196 |
| −0.01 | 2.7320 |
| −0.1 | 2.8680 |

**Table 9–2**

| $x$ | $L(x)$ |
| --- | --- |
| 2.65 | 0.975 |
| 2.66 | 0.978 |
| 2.67 | 0.982 |
| 2.68 | 0.986 |
| 2.69 | 0.990 |
| 2.70 | 0.993 |
| 2.71 | 0.997 |
| 2.72 | 1.001 |
| 2.73 | 1.004 |
| 2.74 | 1.008 |
| 2.75 | 1.012 |

"I remember having to calculate a decimal approximation for $\pi$," Recordis added. "We came up with $\pi = 3.141592654$. The professor said that we could calculate $\pi$ out to thousands of decimal places if we wanted to, but I made it clear that we were not going to calculate it out to thousands of places unless someone came up with a thousand very good reasons why we should."

"We decided that we had discovered a fundamental irrational number," the professor said. "I had thought $\pi$ would be the only one. We'll have to add the number $e$ to our list."

Recordis entered in his giant record book:

$$e = 2.718 \ldots$$

"Don't get carried away," Trigonometeris warned. "Remember that we still have to get the values of the $L(x)$ function from Builder. If you guys are right and $L(x) = \log_e x = \log_{2.718} x$, then $L(e) = L(2.718)$ must equal 1. If Builder comes back with a different value, we're sunk."

We waited nervously for Builder to come back with his table of values. Nobody wanted to rush him because we knew that he must do an accurate job. Finally there was a knock at the door, and Builder came in. "This table should have all the values you could possibly want," he said, exhausted.

"Read off all the values between 2.65 and 2.75," the professor requested. Builder looked puzzled, because these seemed like strange numbers, but he read them off (Table 9–2).

"It does work!" the professor exclaimed. "$L(x)$ does equal 1 somewhere between 2.71 and 2.72, just as it should. Then $L(x)$ must be the logarithm function."

"We had better write this as a rule," Recordis said. "First we need to think of a short way to write this function. We wrote $\log x$ to stand for $\log_{10} x$, so let's think of something to stand for $\log_e x$."

"We could write $\log_e x = \ln x$," the king offered. "I don't know why, but it looks pretty."

"That's good because it has fewer letters," Recordis said. "And we need to think of a name for a logarithm to the base $e$." I suggested the name *natural logarithm,* since it seemed as though this function had arisen in the natural course of our investigations of the properties of integrals.

## NATURAL LOGARITHM OF A NUMBER

natural logarithm of $x = \ln x = \log_e x = \int_1^x \frac{1}{t}\, dt$

(where $e = 2.718$)

---

### FIRST AMENDMENT TO THE POWER RULE FOR INTEGRALS

$$\int x^n \, dx = \frac{1}{n+1} x^{n+1} + C \qquad \text{if } n \neq -1$$

$$= \ln |x| + C \qquad \text{if } n = -1$$

---

(We later investigated what happens if $x$ is negative. The logarithm of a negative number is not defined, but we can still use the logarithm function in the power rule if we take the absolute value of $x$ ($|x|$).)

"We could also write that rule in terms of definite integrals," the professor said.

$$\int_a^b x^{-1} \, dx = \ln b - \ln a$$

"No," Recordis said firmly. "We know that the connection between definite integrals and indefinite integrals always holds, so we definitely do not need to write each rule once in terms of indefinite integrals and once again in terms of definite integrals."

We went back to Count Q's problem:

$t = -4 \int x^{-1} \, dx$, with the initial condition $x = 4$ million when $t = 0$

"We can solve that right away," the professor said.

$$t = -4 \ln x + C$$

"Using the initial condition gives us the following expressions."

$$0 = -4 \ln(4 \text{ million}) + C$$

$$C = 4 \ln(4 \text{ million})$$

"Builder, can you give us a value for ln(4 million)?" the king asked.

Builder looked aghast, but a few minutes later he came back with the answer from his logarithm pools.

$$\ln(4 \text{ million}) = 15.2$$

$$C = 4 \times 15.2 = 60.8$$

"Now what does $t$ equal when $x$ is zero?" Recordis said. "That will tell us how long it will take Rutherford to scoop up all the beads."

$$t = -4 \ln x + 60.8$$

"Hold it," the professor protested. "We said that there was no such thing as $L(0)$ or $\log_{(\text{any base})} 0$. That means ln 0 doesn't exist."

"That means Rutherford can *never* pick up all the beads!" Trigonometeris said in dismay.

"Of course, that's right," the king stated. "The count said that in any time period Rutherford could scoop up only one fourth of the beads that were left. That way he could never get them all. Even if there was only 1 gram of beads left, he could pick up only one fourth of that gram."

"Don't be so picky!" Recordis said disrespectfully. "If Rutherford can sweep up all but 1 gram of beads, I will be glad to personally pick up the rest."

We easily calculated the value of $t$ for $x = 1$:

$$t = -4 \ln 1 + 60.8 = 60.8 \text{ hours}$$

"That's $2\frac{1}{2}$ days," Recordis said. "Not bad, considering how many beads there were to start with." Recordis went running off to take the answer to the count. Sure enough, the count did give us a nice gift for providing him with the answer. We were still amazed at the strange paths along which this investigation was taking us.

## Note to Chapter 9

If you look carefully at a definite integral, you will see that the name of the variable of integration (the variable in the $d$-variable term) does not make any difference to the final value of the integral. For example, the $q$ in the integral defining the logarithm function could have been renamed $g$ or $q'$ or $s''$ without affecting the value of $L(x)$:

$$L(x) = \int_1^x \frac{1}{q} \, dq = \int_1^x \frac{1}{g} \, dg = \int_1^x \frac{1}{q'} \, dq' = \int_1^x \frac{1}{s''} \, ds''$$

The variable of integration in a definite integral is sometimes called a *dummy variable*.

Notice that for an indefinite integral, however, the name of the variable of integration does matter:

$$\int x^2\, dx = \tfrac{1}{3}x^3 + C \neq \int y^2\, dy = \tfrac{1}{3}y^3 + C$$

## Exercises

1. On one occasion the gremlin tried to take over the kingdom with a bacteria dish. The bacteria multiplied at such a rate that $dn/dt = 3n$. At $t = 0$ hour, the number of bacteria was 10. If the gremlin had not been stopped, how long would it have been before $n$ equaled 1000?

Find $y'$ for:

2. $y = \ln x^2$
3. $y = \ln(-x)$
4. $y = \ln \sin x$
5. $y = \ln \tan x$
6. $y = \ln(x^2 + 4)^{1/2}$
7. $y = \ln(ax^2 + bx + c)$
8. $y = \ln(x\sqrt{x + 1})$
9. Use the definition of the derivative to find $y'$ for $y = \log_{10} x$.
10. Sketch the graph of the curve $y = (\ln x)/x$. Find any points with horizontal tangents. Find any points of inflection.
11. Use 10 inscribed rectangles to estimate ln 2. (For a helpful table of values, see Table 9–3.)

**Table 9–3**

| $x$ | $1/x$ |
|-----|-------|
| 1.1 | 0.9091 |
| 1.2 | 0.8333 |
| 1.3 | 0.7692 |
| 1.4 | 0.7143 |
| 1.5 | 0.6667 |
| 1.6 | 0.6250 |
| 1.7 | 0.5882 |
| 1.8 | 0.5556 |
| 1.9 | 0.5263 |
| 2.0 | 0.5000 |

12. For $\ln x = \displaystyle\int_1^x (1/t)\, dt$, show that $\ln(a/b) = \ln a - \ln b$.

13. For $\ln x = \displaystyle\int_1^x (1/t)\, dt$, show that $\ln(x^n) = n \ln x$.

Evaluate:

**14.** $y = \int \dfrac{1}{x + 4}\, dx$

**15.** $y = \int \dfrac{1}{3x + 4}\, dx$

**16.** $\int \dfrac{1}{x^2 + 2x + 1}\, dx$

**17.** $\int \dfrac{1}{ax + b}\, dx$

**18.** $\int \dfrac{1}{(ax + b)^2}\, dx$

**19.** $\int \dfrac{x}{3x^2 + 6}\, dx$

**20.** $\int \dfrac{2x + 4}{x^2 + 4x + 6}\, dx$

**21.** $\int \dfrac{\sin \theta}{\cos \theta}\, d\theta$

**22.** $\int \dfrac{\cos \theta}{\sin \theta + 4}\, d\theta$

**23.** Consider the function $y = f(x) = (1 + x)^{1/x}$. Find $f(x)$ for the following values of $x$: 10, 8, 6, 4, 2, 1, 0.5, 0.25, 0.1, $-0.1$, $-0.25$, $-0.5$, $-0.6$, $-0.7$, $-0.8$, $-0.9$. Make a sketch of the curve. What is $f(0)$? What can you say about $\lim_{x \to 0} f(x)$? What is $\lim_{x \to \infty} f(x)$?

**24.** The *work* done in compressing a container of gas is given by $-\int_{V_1}^{V_2} P \, dV$, where $P$ is the pressure of the gas, $V_1$ is the initial volume, and $V_2$ is the final volume. For a gas that obeys the ideal gas law ($PV = $ constant), find the work done in compressing the gas from $V_1$ to $V_2$. What is the work if $V_1 = 2V_2$? What is the sign of the work if $V_1 > V_2$?

# 10
# Exponential Functions and Integration by Parts

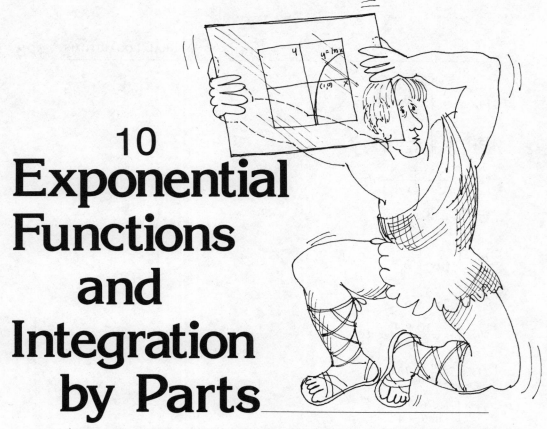

Mongol decided that the logarithm pool made a perfect adjustable swimming pool, and he had fun splashing around in the logarithms of different numbers. Meanwhile, Builder had finished the table of logarithms, and now we had to decide what to do with it.

"Maybe we should make a graph of the logarithm function," Trigonometeris suggested. "After we first discovered the trigonometric functions, we found that it helped to make graphs of them."

Builder got a giant glass plate out of his supply room and brought out his best etching equipment.

"The easiest point is $x = 1$," the king said. "We know that $y = \ln 1 = 0$."

"We also know that $\ln x$ is not defined for negative numbers," the professor added. "That means that we need to worry only about points that are to the right of the line $x = 0$."

"We can use what we did with derivatives," Recordis said. "Remember concave up and concave down?"

The derivatives were easy to calculate:

$$y = \ln x$$

$$y' = \frac{1}{x}$$

$$y'' = (-1)x^{-2} = \frac{-1}{x^2}$$

124

"We know that $y'$ is never zero," the king said. "This means that the curve must never have a horizontal tangent."

"And $y'$ is always positive, so the curve is sloping upward all the time."

"And $y''$ is always negative," Recordis said. "This means that the curve is always concave downward."

With this information, plus the results from the table of values, Builder was able to etch the curve on the glass plate. (See Figure 10–1.)

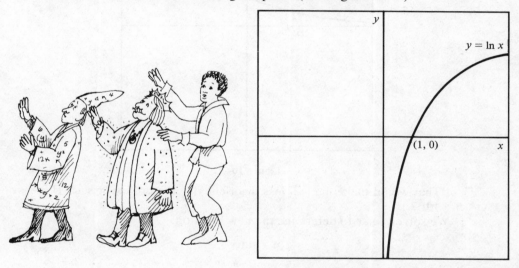

**Figure 10–1.**

"Let's store this graph in the National Archives Room," the king suggested. Nobody else could carry the large plate, so Mongol began to lift it. The glass was slippery at the edges, and he had barely picked it up when he started to slip. As he fell to the floor of the Main Conference Room, he tried to protect the glass plate. He was able to save it from breaking, but it landed on the ground backwards from the way it had been.

"Not fun!" Mongol cried.

"Mongol, what did you do to the graph?" the professor asked, looking at the backward view of the logarithm function through the glass plate. Mongol began to cry, thinking that the professor was terribly angry at him.

"You made a whole new graph!" the professor said. "I've never seen a graph like this before!" She began to get excited, and she congratulated Mongol. Mongol smiled and jumped up and down.

"We have created an *inverse function,*" Recordis said, after the kingdom had stopped shaking. "A long time ago we said that we could do that if we traded the $x$ and $y$ coordinates on a graph."

"Let's find out what the inverse function is," the king suggested. "We still have $y = \ln x$. Let's trade $x$ and $y$ around so that the horizontal axis is $x$ again and the vertical axis is $y$ again."

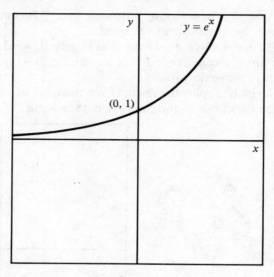

**Figure 10–2.**

"That would make me feel less disoriented," the professor said. (Figure 10–2.)

We solved for $y$ to determine the new function:

$$x = \ln y$$

$$e^x = e^{\ln y}$$

$$y = e^x$$

"Let's call it an *exponential function*," the professor said, "since we're raising $x$ to an exponent."

"Is it good for anything?" Recordis asked.

"I think it's a good way to get big numbers fast," the professor replied thoughtfully. "Recordis, suppose I gave you a choice between having, after $x$ days, either $x^{20}$ or $e^x$ dollars. Which would you choose?" Recordis, who knew something about algebraic functions, chose $x^{20}$, so the professor was left with $e^x$. We made a table of the number of dollars each would have on a given day.

| $x$ (days) | $x^{20}$ (Recordis) | $e^x$ (professor) |
|---|---|---|
| 1 | 1 | 2.7 |
| 2 | 1,048,576 | 7.4 |
| 3 | 3.5 billion | 20.1 |
| 4 | 1.1 trillion | 54.6 |
| 5 | $9.5 \times 10^{13}$ | 148 |

"It looks as though you have no chance," Recordis said, feeling sorry for the professor. "Don't worry. I'll be generous."

"Let's wait a bit longer," the professor said.

| 10 | $1.0 \times 10^{20}$ | $2.2 \times 10^4$ |
| 20 | $1.0 \times 10^{26}$ | $4.9 \times 10^8$ |
| 50 | $9.5 \times 10^{33}$ | $5.2 \times 10^{21}$ |

"The professor is catching up!" the king remarked.

"Still, if I had that kind of money I could buy the whole world!" Recordis said.

| 100 | $1.0 \times 10^{40}$ | $2.7 \times 10^{43}$ |
| 200 | $1.0 \times 10^{46}$ | $7.2 \times 10^{86}$ |

"The professor was right!" the king said, surprised. "In less than a year she would have $7.2 \times 10^{40}$ = 72,000,000,000,000,000,000,000,000, 000,000,000,000,000 times as much money as Recordis would."

"I still wouldn't mind having $1.0 \times 10^{46}$ dollars," Recordis replied defensively. "Besides, how do we know that the function $e^x$ doesn't turn around and start to get smaller for larger values of $x$?"

"We can check the derivative and see whether it is ever negative," the professor said.

We used the power rule:

$$y = e^x$$

$$y' = xe^{x-1}$$

"That was easy," the professor said.

"It doesn't work, though," Recordis pointed out, anxious to prove that he wasn't as dumb as the professor had just made him look. "Look at $x = 0$."

$$x = 0$$

$$\left( \text{King's idea of } \frac{dy}{dx} \right) = 0$$

"But if you look at the graph (Figure 10–2) it is clear that when $x = 0$ the slope of the curve is not zero."

"What happened?" the professor said. We were all stunned. "The power rule always worked before."

Finally the king had a suggestion. "We proved the power rule using an exponent that was a constant. Maybe it doesn't work if you have a constant number raised to a variable exponent."

We tried to find the derivative of $e^x$ directly, using the definition of the derivative, but a few minutes later we gave up. Finally Recordis came up with a suggestion.

"I remember some of the work we did with logarithms," he said. "We found that they were most useful because they could make some calculations a lot simpler. Naturally, I liked them very much. It seemed to me that, whenever we had numbers raised to powers, it helped to take the logarithm of both sides of the equation.''

$$y = e^x$$

$$\ln y = \ln e^x = x$$

"Now we don't have an explicit function," the king said, "but maybe we can use implicit differentiation." (See Chapter 4.)

We applied $d/dx$ to both sides:

$$\frac{d}{dx} \ln y = \frac{d}{dx} x$$

We could differentiate the left-hand side using the chain rule:

$$\frac{1}{y} \frac{dy}{dx} = 1$$

$$\frac{dy}{dx} = y$$

$$= e^x$$

"That can't be right!" the professor objected. "You can't have a function whose derivative is itself!"

"Why not?" Recordis asked. The professor couldn't think of any reason why a function couldn't have a derivative equal to itself.

"That means you could take the second derivative and get the same thing!" the king said.

$$y = e^x$$

$$\frac{d^2y}{dx^2} = e^x$$

"You could take the hundredth derivative and still get the same thing!" Recordis exclaimed.

$$y = e^x$$

$$\frac{d^{100}y}{dx^{100}} = e^x$$

"Amazing!" the professor murmured. "The function is indestructible! You could differentiate it forever and still not change it!"

"We can even work that backwards," Recordis said.

$$\int e^x \, dx = e^x + C$$

"Let's see what happens if we raise some other number to the $x$ power," the professor suggested.

We used logarithms to find the derivative of $y = 2^x$:

$$\ln y = \ln 2^x = x \ln 2$$

$$\frac{1}{y}\frac{dy}{dx} = \ln 2$$

$$\frac{dy}{dx} = 2^x \ln 2$$

"At least its derivative isn't equal to itself," the professor remarked. We went through exactly the same calculations to make a general rule:

---

## DIFFERENTIATION RULE FOR EXPONENTIAL FUNCTIONS

$$y = a^x$$

$$\frac{dy}{dx} = (\ln a)a^x$$

---

"What if we have a variable in both the exponent and the base?" the professor asked. "How about $y = x^x$?" We used the same method, and it was only a little more complicated this time.

$$y = x^x$$

$$\ln y = x \ln x$$

$$\frac{1}{y}\frac{dy}{dx} = x\frac{1}{x} + \ln x$$

$$\frac{dy}{dx} = x^x + (\ln x)x^x$$

"We should remember this method," Recordis said, always looking for simpler ways to do things. "We'll call it the method of *logarithmic implicit differentiation*."

"What other kinds of functions would you use it for?" the king asked.

"I have been having nightmares about complicated functions that I was afraid you would have me differentiate some day. Naturally, once you people have developed all the theory, I get stuck with all the hard work. Just imagine a function like . . . ," and Recordis wrote down the first complicated function that came into his head.

$$y = \sqrt{\frac{(x-1)(x+3)^2(x+4)^3}{(x+1)(x+2)(x-3)}}$$

"We could find *dy/dx* if we had to," the professor said, "using the power rule, the chain rule, and the product rule."

"But it will be a lot easier like this," Recordis said. "Look what happens if we take the logarithm of both sides."

$$\ln y = \ln\left[\frac{(x-1)(x+3)^2(x+4)^3}{(x+1)(x+2)(x-3)}\right]^{1/2}$$

Using the properties of logarithms that we knew, Recordis was able to simplify this expression quite a bit.

$$\ln y = \tfrac{1}{2}\{\ln[(x-1)(x+3)^2(x+4)^3] - \ln[(x+1)(x+2)(x-3)]\}$$

$$= \tfrac{1}{2}[\ln(x-1) + 2\ln(x+3) + 3\ln(x+4) - \ln(x+1)$$
$$- \ln(x+2) - \ln(x-3)]$$

Now we applied $d/dx$ to both sides:

$$\frac{1}{y}\frac{dy}{dx} = \frac{1}{2}\frac{d}{dx}[\ln(x-1) + 2\ln(x+3) + 3\ln(x+4) - \ln(x+1)$$
$$- \ln(x+2) - \ln(x-3)]$$

$$= \frac{1}{2}\left(\frac{1}{x-1} + \frac{2}{x+3} + \frac{3}{x+4} - \frac{1}{x+1} - \frac{1}{x+2} - \frac{1}{x-3}\right)$$

$$\frac{dy}{dx} = \frac{1}{2}\sqrt{\frac{(x-1)(x+3)^2(x+4)^3}{(x+1)(x+2)(x-3)}}$$

$$\left(\frac{1}{x-1} + \frac{2}{x+3} + \frac{3}{x+4} - \frac{1}{x+1} - \frac{1}{x+2} - \frac{1}{x-3}\right)$$

"That's still a complicated answer," the professor said.

"Of course the answer is just as complicated as it would be by any other method!" Recordis retorted. "It is just that the method involved much less work that we would have had if we had tried to find the answer by brute force. We definitely will remember this method when we have to differentiate complicated expressions involving powers or products. Now all I need is a logarithmic curve pool," Recordis went on excitedly. "After all, that is the function we started with before Mongol inverted our graph."

$$y = \ln x \qquad \text{from } x = 1 \text{ to } x = 5$$

"That will give us an expression for the area."

$$\int_1^5 \ln x \; dx =$$

Suddenly he stopped short.

"We don't know how to integrate the logarithm function," Recordis said slowly. His enthusiasm began to fade.

"Why did you decide to make a pool like that, anyway?" the professor asked him. "What if the gremlin comes back and repeats his fire-and-water threat, only we don't know the antiderivative function?"

We pondered for a long time, but we couldn't think of a function that would work.

"Maybe there isn't any function that works," Trigonometeris said.

"That's what we said last time," the king reminded him.

"But then we created a Mysterious Function to represent the area under the $1/x$ function, and we were able to find some simple properties. Suppose we create a Mysterious Function for the area under a logarithm curve, and it doesn't have any simple properties?" the professor asked.

"We can have Builder make another pool and get a table of values," the king said.

"But that could go on forever!" Recordis complained. "We'll make a pool for the integral of the logarithm, and then we'll need a pool for the integral of the integral of the logarithm and then a pool for the integral of the integral of the integral . . ."

"I see your point," the professor said. "We have developed rules that allow us to differentiate any function. But we still haven't developed rules that will allow us to integrate any function."

"That means we just need more integration methods," the king stated confidently.

"But how do we know that we will always be able to find rules that work? Suppose there are some functions that just don't have any kind of simple antiderivative. Then the only way we could integrate them would be to build a pool and measure the area directly."

"Maybe there are lonely functions like that—functions that lack anti-derivatives," the king said. "But I'm sure a nice, simple function like ln $x$ can't be one of them."

There was a long silence before the professor began to develop a carefully planned idea. She was anxious to prove that she was the one who came up with the ideas when they really counted.

"The only way to write an integration rule is to rewrite a differentiation rule backwards," she said slowly. "We did that for the power rule, and we did approximately the same thing for the differentiation chain rule when we developed the method of integration by substitution. But . . . we still have not developed an integral form for the differentiation product rule!"

Igor wrote down the rule:

$$\frac{d(uv)}{dx} = u\,\frac{dv}{dx} + v\,\frac{du}{dx}$$

"Remember what we said about differentials," Recordis reminded her. "We can simplify both sides by multiplying them by $dx$."

$$d(uv) = u\,dv + v\,du$$

"Let's integrate," the professor said.

$$\int d(uv) = \int u\ dv + \int v\ du$$

$$uv = \int u\ dv + \int v\ du$$

$$\int u\ dv = uv - \int v\ du$$

"How does that help?" the king asked.

"Why, $u$ and $v$ can be anything!" the professor said. "Once we define $u$ and $v$, we can rewrite any integral we want in this form!"

"So what?" Recordis asked. "We still haven't gotten rid of the integral sign! We still have to evaluate the integral $\int v\ du$."

"But that integral might be simpler!" the professor said.

"But it might be more complicated!" Recordis retorted.

"Well, it *might* be simpler!" the professor said.

"There is only one way to decide," the king interrupted. "If this method results in a simpler integral, then we will use it. If it results in a more complicated integral, then we won't."

We went back to our problem: $\int \ln x\ dx$.

"Let's try these definitions," the professor said.

$$\text{Let } u = \ln x.$$

$$\text{Let } dv = dx.$$

"That's all right," Recordis said. "You're just changing the name of a variable. But now you need to find $du$ and $v$."

$$u = \ln x$$

$$du = \frac{dx}{x}$$

$$dv = dx$$

$$v = x$$

We put these values into the professor's equation to see whether we ended up with a simpler integral:

$$\int u\ dv = uv - \int v\ du$$

$$\int \ln x\ dx = (\ln x)(x) - \int x\ \frac{1}{x}\ dx$$

$$= x \ln x - \int dx$$

"It works!" the professor said. "We did end up with a second integral that is much simpler than the first. We have found a new integration method!"

$$\int \ln x\ dx = x \ln x - x + C$$

"Let's differentiate this expression and make sure it works!" Recordis urged, before the professor got carried away.

$$\frac{d}{dx}(x \ln x - x + C) = ? \text{ (had better be } \ln x)$$

$$= \frac{d}{dx}(x \ln x) - \frac{dx}{dx} + \frac{dC}{dx}$$

$$= x\frac{d}{dx}\ln x + \ln x\frac{dx}{dx} - 1 + 0$$

$$= x\frac{1}{x} + \ln x - 1$$

$$\frac{d}{dx}(x \ln x - x + C) = \ln x$$

"It does work!" Recordis said. "We can find the area of the pool now."

$$A = \int_1^5 \ln x\, dx = (x \ln x - x)\bigg|_1^5 = 5\ln 5 - 5 - (1\ln 1 - 1)$$

$$= 4.047$$

"What should we call this method?" Recordis asked.

"We'll call it the method of *integration by parts*," the king suggested. "The whole idea is to break the integral up into two parts: $u$ and $dv$."

---

## METHOD OF INTEGRATION BY PARTS

When an integral defies any other means of solution, split it into two parts: call one part $u$, and the other part $dv$ (which must include the differential—the $d$-variable term). Then differentiate $u$ to obtain $du$, and integrate $dv$ to get $v$. Then use the formula:

$$\int u\, dv = uv - \int v\, du$$

If $\int v\, du$ looks simpler than the original integral ($\int u\, dv$), you are making progress and can proceed to a solution. If the integral $\int v\, du$ looks more complicated than the original integral, you are probably hopelessly lost and should either (a) choose new values for $u$ and $dv$, (b) try another method, or (c) give up.

---

"Let's try a weird integral and see whether we can do it," the professor said. "How about $\int x \sin x\, dx$?"

"You don't want to do that!" Recordis objected. "That first $x$ is a variable, rather than a constant, so you can't move it outside the integral sign."

We tried integration by parts:

$\int x \sin x \, dx$

Let $u = x \sin x$.

Let $dv = dx$.

$du = (x \cos x + \sin x) \, dx$

$v = x$

$\int u \, dv = uv - \int v \, du$

$\int x \sin x \, dx = (x \sin x)(x) - \int (x)(x \cos x + \sin x) \, dx$

"That did not help!" Recordis said. "The new integral we came up with is *not* simpler than the original integral!"

The professor was stumped, but the king suggested, "Let's redefine $u$ and $dv$."

Let $u = x$.

Let $dv = \sin x \, dx$.

$du = dx$

$v = -\cos x$

Putting that into the formula, we found that:

$$\int x \sin x \, dx = (x)(-\cos x) - \int (-\cos x) \, dx$$
$$= -x \cos x + \sin x + C$$

"That's the answer!" the king exclaimed. We were able easily to differentiate this result to prove that the mysterious method of integration by parts had indeed produced the correct answer.

"I was wondering when we would come across a last-resort method like this," Recordis said. "I bet we can integrate almost anything!"

"Don't be so hasty!" the professor warned, afraid that the gremlin might be spying. Still, we were all in a happy mood as this time we were able to put the glass plate of the logarithm function safely into the archives, where it would stand as a monument to the progress we had made.

## Note to Chapter 10

It turns out that the exponential function will always increase faster than any polynomial function for large values of $x$. This holds true no matter how big an exponent you use; for example, for large enough $x$, $e^x$ is even bigger than $x^{100,000}$. Stating this as a theorem, we have

$$\lim_{x \to \infty} \frac{x^a}{e^x} = 0 \qquad \text{for any } a, \text{ no matter how large}$$

Many functions lack antiderivatives that can be expressed in terms of elementary functions. The only way to integrate such a function is basically the same way in our country as it is in Carmorra. The method is known as *numerical integration*. Of course, in our country we would use a computer rather than construct a pool to measure the area.

---

_____ Exercises

Find $dy/dx$ for:

**1.** $y = e^{ax}$

**2.** $y = e^{-x^2}$

**3.** $y = e^{ax^2+b}$

**4.** $y = (e^{ax})^m$

**5.** $y = 10^x$

**6.** $y = e^{-ax} \sin x$

**7.** $y = a^{(x^x)}$

**8.** $y = \sqrt{(x - a)(x - b)}$

**9.** $y = \dfrac{x + 1}{(ax^2 + bx + c)^{3/2}}$

**10.** $y = \dfrac{x - 5}{(x + 2)(x + 3)}$

**11.** $y = \dfrac{1}{(2\pi)^{1/2}} e^{-[(x-\mu)^2/\sigma^2]}$

**12.** Use logarithmic implicit differentiation to verify the differentiation power rule.

**13.** Use logarithmic implicit differentiation to verify the differentiation product rule.

**14.** Use the definition of the derivative to show that $(d/dx) e^x = e^x$. Use a fact that we found later:

$$e^x = 1 + x + \frac{x^2}{2!} + \frac{x^3}{3!} + \cdots + \frac{x^n}{n!} + \cdots$$

Evaluate:

**15.** $y = \int e^{ax}\, dx$

**16.** $y = \int e^{ax+b}\, dx$

**17.** $y = \int xe^{x^2}\, dx$

**18.** $y = \displaystyle\int \frac{e^x}{e^x + 5}\, dx$

**19.** $y = \displaystyle\int \frac{\ln x}{x}\, dx$

**20.** Find $dy/dx$ for $y = \sin x - x \cos x$ to verify the integration result from the chapter.

21. Find the area of Recordis' new exponential pool, which is bounded by $y = e^x$, $y = 0$, $x = 0$, and $x = 3$.
22. Find $dy/dx$ for $y = x^2 e^{-x}$. Find points where the curve has horizontal tangents. Make a sketch of the curve.
23. The integral $y = \int \sin x \cos x\ dx$ may be evaluated three ways. (a) Find $y$ by using integration by parts. (b) Find $y$ by making the substitution $u = \sin x$. (c) Find $y$ by using a trigonometric identity to simplify $\sin x \cos x$. Which method is easiest? Is the answer the same for all three methods?
24. Find out how many bacteria there will be at time $t$ in the gremlin's bacteria dish ($dn/dt = 3n$; $t = 0$ when $n = 10$). (Remember that $t$ is measured in hours.) How many bacteria will there be after 5 hours?
25. Solve for $y$: $y = \int x \ln x\ dx$.
26. Verify the above result by differentiation.
27. Solve for $y$: $y = \int xe^x\ dx$.
28. Verify the above result by differentiation.
29. Solve for $y$: $y = \int x^2 e^x\ dx$.
30. Verify the above result by differentiation.
31. Solve for $y$: $y = \int x^2 \ln x\ dx$.
32. Solve for $y$: $y = \int x \sqrt{1 + x}\ dx$.
33. Solve for $y$: $y = \int x^2 \sin x\ dx$.
34. Use integration by parts twice to evaluate $y = \int e^x \cos x\ dx$.
35. A *reduction formula* is a formula that tells how to express a complicated integral in terms of an integral that is slightly simpler. Use integration by parts to derive the reduction formula:

$$y = \int \sin^m x\ dx = -\frac{\sin^{m-1} x \cos x}{m} + \frac{m-1}{m} \int \sin^{m-2} x\ dx.$$

# 11
# Integration
# by
# Trigonometric
# Substitution

"We should have a party!" Recordis said the next day. "Let's do something to celebrate our escape from the gremlin and all the other good events that have happened lately. We can invite all the children." The king agreed, so we quickly began to make plans for flowers, ribbons, refreshments, and rides.

The only person not enjoying the preparations was Trigonometeris, who was feeling left out again. I wanted to cheer him up, so I suggested that we join Recordis at the office of the Differentiation and Integration Business to see whether anything interesting was happening.

Recordis was talking to the Royal Gardener and National Park Superintendent. "We are planning flowers for the king's rose garden," the gardener was saying. "We want to fill the garden with roses that will bloom on the day of the party. The garden is shaped like an ellipse with length 20 units and width 10 units."

"The king always did like ellipses," Recordis said. "In this case it looks as though the ellipse has a semimajor axis of $a = 10$ and a semiminor axis of $b = 5$."

"We need to know how many rose bushes will be required to fill the garden," the gardener said. "And in order to do that, we need to know the area of the ellipse. Can you tell us what the area of the ellipse is?"

"You came to the right place," Recordis said, only this time he was not as overconfident as he had been when he was visited by his neighbor, Count Q. "We set up what we call a definite integral." He riffled through his notes to find the equation for an ellipse (which he had forgotten again).

---

## EQUATION OF ELLIPSE WITH CENTER AT ORIGIN

$$\frac{x^2}{a^2} + \frac{y^2}{b^2} = 1$$

(where $a$ = semimajor axis, $b$ = semiminor axis)

---

"We can solve for $y$ in terms of $x$, using algebra," Recordis said.

$$\frac{y^2}{b^2} = 1 - \frac{x^2}{a^2}$$

$$y^2 = b^2\left(1 - \frac{x^2}{a^2}\right)$$

$$y = \pm b\sqrt{1 - x^2/a^2}$$

We decided that we would first find the area of the quarter of the ellipse where $x$ and $y$ are both positive, so we used the plus sign in front of the square root.

$$y = b\sqrt{1 - x^2/a^2}$$

"Is that the answer?" asked the gardener, who was getting impatient.
"No," Recordis said. "Now we set up the definite integral."

$$\text{(area of quarter-ellipse)} = b\int_0^a \sqrt{1 - x^2/a^2}\, dx$$

$$\text{(area of whole ellipse)} = 4b\int_0^a \sqrt{1 - x^2/a^2}\, dx$$

"Now, if you'll excuse us, we'll be right back after we figure out how to do the integral," Recordis said, and before the gardener knew what was happening Recordis had closed the business and was running through the palace shouting, "Help!"

"Do you always have problems?" the professor asked as we gathered in the Main Conference Room a few minutes later.

"This time it's the king's garden that is causing the problem," Recordis said. He had Igor display the integral that we didn't know how to do. "We can't use the power rule, because we don't have $\int x^n\, dx$. And we can't substitute like this: Let $u = 1 - x^2/a^2$, because then we would have $du = -2x\, dx/a^2$, and we don't have any way to turn $dx$ into $du$ (since we can't take that $x$ outside the integral sign)."

"We could solve it if we had an extra $x$ outside the square root sign," the king said. "Suppose it was written as follows."

$$z = 4b \int_0^a x \sqrt{1 - x^2/a^2} \, dx$$

"Then we could substitute."

$$u = 1 - \frac{x^2}{a^2}, \qquad du = \frac{-2x \, dx}{a^2}$$

"And," he continued, "we could turn $dx$ into $du$ by doing this."

$$z = 4b \int_{u=1}^{u=0} u^{1/2} \left( \frac{-a^2}{2} \right) \left( \frac{-2}{a^2} \right) x \, dx$$

$$= 4b \int_1^0 u^{1/2} \left( \frac{-a^2}{2} \right) du$$

$$= -2ba^2 \, \tfrac{2}{3} \, u^{3/2} \, \Big|_1^0$$

$$z = \frac{4ba^2}{3}$$

"Sure, that was easy," the professor said. "But it doesn't help us, because we have $4b \int_0^a \sqrt{1 - x^2/a^2} \, dx$ without an extra $x$."

"If only we could get rid of the square root sign!" Recordis moaned. "I never did like square root signs anyway."

We stared at the integral for a long time. "We need to make a substitution like this," the professor said.

$$\sqrt{1 - \text{variable}^2} = \text{something simple}$$

"Can anyone think of anything that we have *ever* done that looks like that?"

"I can," Trigonometeris said, suddenly becoming cheerful. "It's obvious. We have $\sqrt{1 - \sin^2 \theta} = \cos \theta$."

"I don't see any need to bring trigonometry into this!" Recordis objected.

"Try this," Trigonometeris said.

$$A = 4b \int_0^a \sqrt{1 - x^2/a^2} \, dx$$

Let $\dfrac{x}{a} = \sin \theta$.

$x = a \sin \theta$

$dx = a \cos \theta \, d\theta$

$\theta = \arcsin\left( \dfrac{x}{a} \right)$

"What's an arcsin?" Recordis asked. "I remember the name, but I forget what it is."

"It's the inverse function for the sine function," Trigonometeris said. "In just the same way, the exponential function is the inverse function for the logarithm function. This is what we said."

If $a = \sin b$, then $b = \arcsin a$.

"That means that $\sin(\arcsin a) = a$ for any $a\,(-1 \leq a \leq 1)$."

"What happens when you put that into the integral?" the king asked. We made the change in the three places where we knew we had to: the integrand, the two limits of integration, and the differential $dx$ term.

$$A = 4b \int_{\arcsin(0/a)}^{\arcsin(a/a)} \sqrt{1 - \sin^2 \theta}\; a \cos \theta\; d\theta$$

"And we know what $\sqrt{1 - \sin^2 \theta}$ is," Trigonometeris said.

"We do?" Recordis asked.

"It's $\cos \theta$!" Trigonometeris answered. "That's what I just told you!"

"Maybe you had better review the list of trigonometric identities, just to make sure Recordis remembers them all," the professor said, trying to conceal the fact that she couldn't remember them all either.

Trigonometeris pulled a page from his book and displayed the following identities:

## TRIGONOMETRIC IDENTITIES (TRUE FOR ALL $A$, $B$)

$\sin^2 A + \cos^2 A = 1$
$\sin^2 A = \frac{1}{2}(1 - \cos 2A)$
$\cos^2 A = \frac{1}{2}(1 + \cos 2A)$
$\sin(-A) = -\sin A$
$\cos(-A) = \cos A$
$\tan A = \sin A/\cos A$
$\text{ctn } A = 1/\tan A$
$\sec A = 1/\cos A$
$\csc A = 1/\sin A$
$1 + \tan^2 A = \sec^2 A$
$1 + \text{ctn}^2 A = \csc^2 A$
$\sin(A + B) = \sin A \cos B + \cos A \sin B$
$\cos(A + B) = \cos A \cos B - \sin A \sin B$
$\tan(A + B) = (\tan A + \tan B)/(1 - \tan A \tan B)$
$\sin 2A = 2 \sin A \cos A$
$\cos 2A = \cos^2 A - \sin^2 A$
$\tan 2A = 2 \tan A/(1 - \tan^2 A)$

We used the identity in the integral:

$$A = 4ba \int_0^{\pi/2} \sqrt{\cos^2 \theta} \cos \theta \, d\theta$$

"Now we can get rid of the square root sign," Recordis said, with sudden excitement. "Maybe this is a good method after all."

$$A = 4ab \int_0^{\pi/2} \cos^2 \theta \, d\theta$$

"How do we do that?" the king asked.
"Use another identity," the professor said.

$$\cos^2 \theta = \tfrac{1}{2}(1 + \cos 2\theta)$$

$$A = 4ab \int_0^{\pi/2} \tfrac{1}{2}(1 + \cos 2\theta) \, d\theta$$

$$= 2ab \int_0^{\pi/2} d\theta + 2ab \int_0^{\pi/2} \cos 2\theta \, d\theta$$

$$= 2\,ab\theta \Big|_0^{\pi/2} + 2ab\,\tfrac{1}{2}\sin 2\theta \Big|_0^{\pi/2}$$

$$= ab\pi - 2ab \times 0 + ab(\sin \pi - \sin 0)$$

$$A = \pi ab$$

"That's a pretty simple answer," Recordis said.
"It makes sense," Trigonometeris told him. "If you have a rectangle (Figure 11–1), its area is $4ab$. It stands to reason that the inscribed ellipse would have an area of about $3ab$."

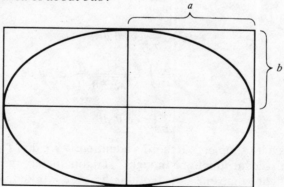

**Figure 11–1.**

"The formula also works for a circle," the professor pointed out. "We said that a circle is the same as an ellipse with $a = b$. In that case the area would be $\pi a^2$, where $a$ is the radius. We already know that that's the correct answer for the area of a circle."

We easily found the area of the rose garden:

$$A = (3.14)(10)(5) = 157$$

"We'll call this method the method of *integration by trigonometric substitution*." Everyone protested that Trigonometeris was naming the method after himself, but I couldn't think of any better name.

"Is the method very generally applicable?" the king asked.

"I'm sure it is," Trigonometeris answered. "I can think of lots of integrals where you would use it."

"Name one," Recordis said.

Trigonometeris gave us the first complicated integral that came into his head:

$$z = \int (a^2 + b^2 x^2)^{-1/2} \, dx$$

"How can you do that?" Recordis asked. "It's an indefinite, rather than a definite, integral, so you can't change the bounds after you make the substitution."

"We can do the same thing," Trigonometeris told him. "At the end we will have to substitute backwards and write $x$ in terms of $\theta$." Trigonometeris thought that the key identity was $\tan^2 \theta + 1 = \sec^2 \theta$. So we made the substitution $bx/a = \tan \theta$:

$$z = \int (a^2 + b^2 x^2)^{-1/2} \, dx$$

$$= \frac{1}{a} \int \frac{1}{\sqrt{1 + b^2 x^2/a^2}} \, dx$$

$$x = \frac{a \tan \theta}{b}$$

$$dx = \frac{a}{b} \sec^2 \theta \, d\theta$$

$$z = \frac{1}{a} \int \frac{1}{\sqrt{\sec^2 \theta}} \frac{a}{b} \sec^2 \theta \, d\theta$$

$$= \frac{1}{b} \int \sec \theta \, d\theta$$

"That is a lot simpler," Recordis admitted. "We don't know how to integrate the secant function, though." Trigonometeris spent a long time trying to figure that one out. Nothing he tried worked. Finally he said, "I'll have the answer tomorrow, I promise."

"I don't suppose the method of trigonometric substitution is much good for any other type of integral," Recordis said. "It surely didn't help much for that one."

"It will help if you have an expression like $(1 + x^2)$ raised to a negative power," Trigonometeris retorted defensively. He was deeply embarrassed that he had been stuck with an integral of one of his own functions that he could not do. "Suppose you had $(1 + x^2)^{-1}$. If you had $(1 + x^2)$ raised to a power that is a positive integer, you could multiply out the result quite easily: $(1 + x^2)^2 = 1 + 2x^2 + x^4$."

"Let's try the other one you mentioned," the professor said.

$$z = \int \frac{1}{1 + x^2} \, dx$$

We decided to use the secant-tangent identity again, so we made the substitution:

Let $x = \tan \theta$.

$\theta = \arctan x$

$dx = \sec^2 \theta \, d\theta$

$$z = \int \frac{1}{1 + \tan^2 \theta} \sec^2 \theta \, d\theta$$

$$= \int \frac{1}{\sec^2 \theta} \sec^2 \theta \, d\theta$$

$$= \int d\theta$$

"That's a perfect integral!" the professor exclaimed.

$$z = \theta$$

"We have to substitute back in for $\theta$," Recordis said.

$$\int \frac{1}{1 + x^2} \, dx = \arctan x + C$$

"That integrates into a standard function!" the king pointed out in surprise.

"But if that's the case, then we can do this," the professor said.

$$\frac{d(\arctan x)}{dx} = \frac{d}{dx} \int \frac{1}{1 + x^2} \, dx = \frac{1}{1 + x^2}$$

Trigonometeris was stunned. "This equation tells us how to differentiate the arctangent function," he said slowly. "How could I have forgotten to look for the derivatives of the inverse functions when we looked for the derivatives of the other trigonometric functions?" He kicked himself (rather hard). "They were right under my nose. How could I have overlooked them?"

We decided to find the derivatives of the other inverse trigonometric functions. After a couple of guesses that didn't work, we tried to evaluate the integral:

$$z = \int \frac{1}{\sqrt{1 - x^2}} \, dx$$

We realized that this type of integral called for the sine-cosine identity:

Let $x = \sin \theta$.

$$\theta = \arcsin x$$

$$dx = \cos \theta \, d\theta$$

$$z = \int \frac{1}{\sqrt{1 - \sin^2 \theta}} \cos \theta \, d\theta$$

$$= \int \frac{\cos \theta \, d\theta}{\cos \theta}$$

$$= \int d\theta$$

$$= \theta$$

$$z = \arcsin x$$

$$\frac{d \arcsin x}{dx} = \frac{1}{\sqrt{1 - x^2}}$$

"That even fits," the king said. "Remember that arcsin $x$ is defined only for $-1 \le x \le 1$, which are the same values of $x$ where $(1 - x^2)^{-1/2}$ is defined."

"The derivative of $y = \arccos x$ should be just the negative of the derivative of $y = \arcsin x$," Recordis guessed. We were able to establish that

$$\frac{d \arccos x}{dx} = - \frac{1}{\sqrt{1 - x^2}}$$

As Recordis started to make a list of the integrals we had done, Trigonometeris got up to leave the room. "Not so fast," Recordis said. "Remember that you still have to tell us how to integrate the secant function. We'll be waiting for your answer tomorrow."

## Exercises

Solve for $y$:

**1.** $y = \displaystyle\int \frac{dx}{x^2 \sqrt{1 + x^2}}$

**2.** $y = \displaystyle\int \frac{dx}{x \sqrt{x^2 - 1}}$

**3.** $y = \displaystyle\int \frac{e^x}{1 + e^{2x}} \, dx$

**4.** Evaluate: $\displaystyle\int_0^{a/b} \sqrt{a^2 - b^2 x^2} \, dx$

5. Evaluate these two expressions: $y_1 = \int x \sqrt{1 - x^2}\, dx$ and $y_2 = x \int \sqrt{1 - x^2}\, dx$. Compare the results.

6. Find the area of a circle of radius $r$.

7. One strip of pink roses will be planted at the tip of the rose garden (Figure 11–2). Find the area of the strip of pink roses.

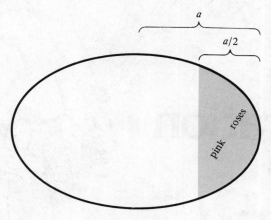

**Figure 11–2.**

8. Evaluate this indefinite integral: $z = 4b \int \sqrt{1 - x^2/a^2}\, dx$. (This is the same as the ellipse area integral.) Verify the result by differentiation.

Solve for $y$:

9. $y = \displaystyle\int \frac{1}{(x/a)^2 + 1}\, dx$

10. $y = \displaystyle\int \frac{1}{x^2 + a^2}\, dx$

11. $y = \displaystyle\int \frac{1}{x^2 + 25}\, dx$

12. $y = \displaystyle\int \frac{1}{x^2 + 4x + 13}\, dx$    (Let $u = x + 2$.)

13. $y = \displaystyle\int \frac{1}{x^2 + 2bx + c}\, dx$    (Let $u = x + b$. Assume that $c > b$.)

14. $y = \displaystyle\int \frac{1}{x^2 + x + 5}\, dx$

15. Solve for $\int \sqrt{1 - x^2}\, dx$ using (a) $x = \sin \theta$ and (b) $x = \cos \theta$. Do the two results agree? Which method is easier?

16. Evaluate $y = \int \arcsin x\, dx$. Verify the result by differentiation.

17. Evaluate $y = \int \arctan x\, dx$. Verify the result by differentiation.

# 12
# Integration by Partial Fractions

"I'd like to see how Trigonometeris weasels his way out of this one," Recordis said when we met the next morning. Trigonometeris had not joined us yet. "Maybe there isn't any way to integrate the secant function. In my opinion, once you've introduced trigonometry to a situation you've created worse problems than you've solved."

The professor wanted to make a summary of all the integration techniques we had developed. She said that we had finished with all the theory that we needed to know for calculus and that all that remained were the menial bookkeeping tasks (meaning that Recordis would have to do the rest of the work). Recordis etched our results on a large plate so that we could have them permanently.

$$\int x^n \, dx = \frac{1}{n+1} x^{n+1} + C \qquad (n \neq -1)$$

$$= \ln |x| + C \qquad (n = -1)$$

$$\int f(u) \frac{du}{dx} \, dx = \int f(u) \, du$$

$$\int \sin x \, dx = -\cos x + C$$

146

$$\int \cos x \, dx = \sin x + C$$

$$\int \tan x \, dx = \ln \left| \sec x \right| + C \text{ (see exercise 9–21)}$$

$$\int \ln x \, dx = x \ln x - x + C$$

$$\int e^x \, dx = e^x + C$$

"We can add what we did yesterday," Recordis said. "We discovered what to do with a quadratic term in the denominator of an integrand." He etched on the plate:

$$\int \frac{1}{1 + x^2} \, dx = ?$$

$$\int \frac{1}{1 - x^2} \, dx = ?$$

He flipped through his notes to find the answer. "This is easy," he said.

$$\int \frac{1}{1 + x^2} \, dx = \arctan x + C$$

Suddenly he turned pale. He stood in front of the last integral that he had written so that we couldn't see it. "That's all we need to do for now," he said.

"Wait a minute," the professor objected. "What about that last one? The one with $\int (1 - x^2)^{-1} \, dx$?"

"What last one?" Recordis said, turning around. "Oh, that last one. It looks as though we didn't do it yesterday, so I think we should forget it."

"But you already wrote it on the plate, so we can't erase it," the king said. "We should try to find out what it is. Let's try this substitution: $x = \sin \theta$, $dx = \cos \theta \, d\theta$."

"I'm warning you: this is only going to cause trouble," Recordis muttered.

$$\int \frac{1}{1 - x^2} \, dx = \int \frac{1}{1 - \sin^2 \theta} \cos \theta \, d\theta$$

$$= \int \frac{\cos \theta}{\cos^2 \theta} \, d\theta$$

$$= \int \frac{1}{\cos \theta} \, d\theta$$

"Oh no!" Recordis moaned.

$$\int \frac{1}{1 - x^2} \, dx = \int \sec \theta \, d\theta$$

"That's exactly what we're trying to find the answer for," the professor said. "I hope Trigonometeris gets here soon with that answer."

"I bet there is no answer, and that's why Trigonometeris isn't here yet,'' Recordis said. ''There must be a way to do this integral without trigonometry. For one thing, we can factor the denominator, since it is the difference of two squares.''

$$\int \frac{1}{1 - x^2}\, dx = \int \frac{1}{(1 - x)(1 + x)}\, dx$$

''Does that help?'' the king asked.

We were trying to figure out what to do with that integral when Builder came in, carrying two bottles of liquids.

''Maybe you can help me,'' he said. ''I'm working on the fireworks for the party. Your work on figuring out the speed of things might help. I have 10 grams of this red juice (call it $x$) and 9 grams of this yellow juice (call it $y$). If you mix them together, they form another liquid, called $z$. It takes exactly 3 grams of $x$ to combine with exactly 2 grams of $y$ to produce exactly 5 grams of $z$. When there is more $x$ and $y$ available, the reaction goes faster; but when there is more $z$ around, the reaction goes slower. In fact, the speed of the reaction is proportional to the product of the amounts of $x$ and $y$ and inversely proportional to the amount of $z$.''

$$\frac{dz}{dt} = \frac{xy}{z}$$

''Can you figure out how much $z$ will have been produced at a particular time after I start mixing them together?''

''We'll at least set up the integral,'' the professor said. We realized that we would have to figure out the amounts of $x$ and $y$ that would be present at a given time, and it was clear that the amounts would be equal to the starting amounts (10 and 9, respectively) minus the amounts that had gone into the formation of the $z$:

$$x = 10 - 0.6z$$

$$y = 9 - 0.4z$$

Putting these expressions into Builder's rate equation, we were able to set up the integral:

$$\frac{dz}{dt} = \frac{xy}{z}$$

$$= \frac{(10 - 0.6z)(9 - 0.4z)}{z}$$

$$\frac{z}{(10 - 0.6z)(9 - 0.4z)}\, dz = dt$$

$$t = \int \frac{z}{(10 - 0.6z)(9 - 0.4z)}\, dz$$

''We can't do that one either,'' the professor said. ''That's basically the same as the other problem we're stuck on. We have two factors in the denominator of a fraction.''

$$\int \frac{1}{(1 - x)(1 + x)} \, dx = ?$$

"We need a new method to tell us what to do when we have a bunch of factors in the denominator like that," the king stated. "If only we had a bunch of fractions *added* together, instead of one giant fraction!"

"It's so easy to take a bunch of fractions that are added together and turn them into one giant fraction," Recordis moaned. "It's too bad you can't do it the other way."

"What did you say was easy?" the professor asked.

"Adding a bunch of fractions together," Recordis said. "Suppose you had this."

$$y = \frac{1}{x - a} + \frac{1}{x - b}$$

"You can easily add these together and turn them into one giant fraction. All you need to do is find the common denominator."

$$y = \frac{1}{x - a} \frac{x - b}{x - b} + \frac{1}{x - b} \frac{x - a}{x - a}$$

$$= \frac{x - b + x - a}{(x - a)(x - b)}$$

$$= \frac{2x - (b + a)}{(x - a)(x - b)}$$

"It's too bad you can't do that procedure backwards," Recordis said. "Then you could start with a fraction having a bunch of factors in the denominator and turn it into a sum of fractions."

"Why can't you do it backwards?" the professor asked.

Recordis didn't say anything for a while. "I thought there was a reason why you couldn't do it backwards," he finally said.

"An equation must work in both directions," the king pointed out. "It may be a little more tricky to start with a single fraction and end up with a sum of fractions, but we should be able to do it. Take, for example, Builder's problem."

$$\frac{z}{(10 - 0.6z)(9 - 0.4z)}$$

"That expression must be the sum of two partial fractions, one with denominator $(10 - 0.6z)$ and one with denominator $(9 - 0.4z)$."

"We don't know what the numerators are, though," Recordis said.

We decided to call the numerators $A$ and $B$ while we tried to solve for them.

$$\frac{z}{(10 - 0.6z)(9 - 0.4z)} = \frac{A}{10 - 0.6z} + \frac{B}{9 - 0.4z}$$

"We don't know what $z$ is," Recordis remarked.

"But this equation must hold for all values of $z$," the king said.
"That means we call it an *identity*," the professor stated.
We found the common denominator for the right-hand side:

$$\frac{z}{(10 - 0.6z)(9 - 0.4z)} = \frac{A(9 - 0.4z) + B(10 - 0.6z)}{(10 - 0.6z)(9 - 0.4z)}$$

"Since these two fractions are equal, their numerators must be equal," Recordis said.

$$z = A(9 - 0.4z) + B(10 - 0.6z)$$
$$= 9A - 0.4Az + 10B - 0.6Bz$$
$$= (-0.4A - 0.6B)z + (9A + 10B)$$

"Now what?" Recordis asked.
"Since this equation must hold for all values of $z$, the coefficient of $z$ on the left-hand side (which is 1) must equal the coefficient of $z$ on the right-hand side, and the constant term on the left-hand side (which is 0) must equal the constant term on the right-hand side." (You can easily verify that, if these two conditions are not met, you can find a value of $z$ such that the equation will not hold.)

coefficients of $z$:

$$1 = -0.4A - 0.6B$$

constant terms:

$$0 = 9A + 10B$$

"That's easy," Recordis said. "That's just two equations in two unknowns."

$$9A = -10B$$

$$A = \frac{-10}{9}B$$

$$1 = -(0.4)\left(\frac{-10}{9}\right)B - 0.6B$$

$$= \frac{4}{9}B - \frac{3}{5}B$$

$$= -\left(\frac{7}{45}\right)B$$

$$B = \frac{-45}{7} = -6.4$$

$$A = \left(\frac{-10}{9}\right)(-6.4) = 7.1$$

"So that means . . ."

$$\frac{z}{(10 - 0.6z)(9 - 0.4z)} = \frac{7.1}{10 - 0.6z} - \frac{6.4}{9 - 0.4z}$$

Recordis was still skeptical, so he double-checked to make sure that the expression on the right did indeed equal the expression on the left. (See exercise 12–6.)

"Now it's easy!" Recordis said. "We can break the original integral into two integrals."

$$t = \int \frac{z}{(10 - 0.6z)(9 - 0.4z)}\, dz = \int \frac{7.1\, dz}{10 - 0.6z} - \int \frac{6.4\, dz}{9 - 0.4z}$$

$$= 7.1 \int (10 - 0.6z)^{-1}\, dz - 6.4 \int (9 - 0.4z)^{-1}\, dz$$

$$= (-7.1)(0.6) \ln|10 - 0.6z| + (6.4)(0.4) \ln|9 - 0.4z| + C$$

$$= -\ln|10 - 0.6z|^{4.3} + \ln|9 - 0.4z|^{2.6} + C$$

$$t = \ln\left|\frac{(9 - 0.4z)^{2.6}}{(10 - 0.6z)^{4.3}}\right| + C$$

Using the initial condition ($t = 0$ when $z = 0$) gives:

$$0 = \ln \frac{9^{2.6}}{10^{4.3}} + C$$

$$= \ln \frac{302.7}{19{,}952} + C$$

$$C = -\ln 0.015$$

$$= 4.2$$

The final answer became:

$$t = \ln \frac{(9 - 0.4z)^{2.6}}{(10 - 0.6z)^{4.3}} + 4.2$$

We realized that it would be pretty hard to solve for $z$ as an explicit function of $t$, but Builder was satisfied when we constructed a table of values (Table 12–1).

"We must think of a name for this method," Recordis said.

"The key seems to be to break the fraction into a sum of partial fractions," the king offered.

"Let's call it the *method of partial fractions*," the professor said.

"Now we can solve this other integral," Recordis said.

$$\int \frac{1}{(1 - x)(1 + x)}\, dx$$

Table 12–1

| $z$ | $t$ |
|---|---|
| 0 | 0.0 |
| 2 | 0.3 |
| 4 | 0.7 |
| 6 | 1.1 |
| 8 | 1.7 |
| 10 | 2.4 |
| 12 | 3.5 |
| 14 | 5.4 |
| 16 | 10.6 |

"There are two factors in the denominator, so that means that there should be two partial fractions, with denominators $(1 - x)$ and $(1 + x)$."

$$\frac{1}{(1 - x)(1 + x)} = \frac{A}{1 - x} + \frac{B}{1 + x}$$

Now it was just a matter of algebra to solve for $A$ and $B$:

$$\frac{1}{(1 - x)(1 + x)} = \frac{A(1 + x)}{(1 - x)(1 + x)} + \frac{B(1 - x)}{(1 - x)(1 + x)}$$

$$\frac{1}{(1 - x)(1 + x)} = \frac{A(1 + x) + B(1 - x)}{(1 - x)(1 + x)}$$

Setting the numerators equal gave us:

$$1 = A + Ax + B - Bx$$

$$= (A - B)x + (A + B)$$

Since the equation must hold for every value of $x$, we could form two equations:

coefficients of $x$:

$$0 = A - B$$

constant terms:

$$1 = A + B$$

These two equations imply:

$$A = B$$

$$1 = A + A$$

$$= 2A$$

$$A = \tfrac{1}{2}$$

$$B = \tfrac{1}{2}$$

"Now we can do the integral easily," Recordis said.

$$\int \frac{1}{(1 - x)(1 + x)} \, dx = \int \frac{1}{2(1 - x)} + \frac{1}{2(1 + x)} \, dx$$

$$= -\tfrac{1}{2} \ln |1 - x| + \tfrac{1}{2} \ln |1 + x| + C$$

$$\int \frac{1}{1 - x^2} \, dx = \tfrac{1}{2} \ln \left| \frac{1 + x}{1 - x} \right| + C$$

At that moment Trigonometeris walked in. He looked terrible, as if he hadn't had any sleep the night before.

"I'll have it for you tomorrow, I promise," he said weakly.

"We don't need you any more," Recordis said unkindly. "We tried a trigonometric substitution on this integral: $\int (1 - x^2)^{-1}\, dx$, and got:

$$x = \sin \theta$$

$$\int \frac{1}{1 - x^2}\, dx = \int \sec \theta\, d\theta$$

"That is exactly the same problem that you haven't been able to find the answer to. But we found another way of getting the answer, using algebra without a trace of trigonometry."

$$\int \frac{1}{1 - x^2}\, dx = \tfrac{1}{2} \ln \left| \frac{1 + x}{1 - x} \right| + C$$

"We called it the *method of partial fractions.*"

"Wait!" the king said. "This means that we can find the integral of the secant function!"

"We can?" Trigonometeris asked.

"Look at this!"

$$x = \sin \theta \rightarrow \int \sec \theta\, d\theta = \int \frac{1}{1 - x^2}\, dx = \tfrac{1}{2} \ln \left| \frac{1 + x}{1 - x} \right| + C$$

"That means we have the following."

$$\int \sec \theta\, d\theta = \tfrac{1}{2} \ln \left| \frac{1 + \sin \theta}{1 - \sin \theta} \right| + C$$

"Of course!" Recordis said. "The old finding-the-integral-of-the-secant-function-by-the-method-of-partial-fractions trick!"

Trigonometeris wanted to contribute something to the discussion, so he found a few ways to simplify this expression for us:

$$\int \sec \theta\, d\theta = \tfrac{1}{2} \ln \left| \frac{(1 + \sin \theta)}{(1 - \sin \theta)} \frac{(1 + \sin \theta)}{(1 + \sin \theta)} \right| + C$$

$$= \tfrac{1}{2} \ln \left| \frac{(1 + \sin \theta)^2}{1 - \sin^2 \theta} \right| + C$$

$$= \tfrac{1}{2} \ln \left| \frac{1 + \sin \theta}{\cos \theta} \right|^2 + C$$

$$= \ln \left| \frac{1 + \sin \theta}{\cos \theta} \right| + C$$

$$= \ln \left| \frac{1}{\cos \theta} + \frac{\sin \theta}{\cos \theta} \right| + C$$

$$\int \sec \theta\, d\theta = \ln \left| \sec \theta + \tan \theta \right| + C$$

"It is amazing that we could get an answer so complicated, yet so simple," the king remarked.

"We must make a general plan for the method of partial fractions," the professor said. "This method will help whenever we have two polynomials in a fraction, such as $N(x)/D(x)$. Here $N$ stands for numerator and $D$ stands for denominator. If we want to, we can say that the *degree* of $N(x)$ is less than the degree of $D(x)$. That's what we used to call a *proper rational function*."

"Why can we say that?" Trigonometeris asked.

"If there was a higher power of $x$ in the numerator than in the denominator, we could use algebraic division to express the fraction as the sum of a polynomial and a proper fraction," the professor answered. "I always thought it was fascinating the way algebraic division worked."

"It is fascinating the way algebraic division works, but it is a *lot* of work," Recordis said. "Anyway, we know that we can factor $D(x)$ into a bunch of linear factors. Then we know that each of those linear factors will be in the denominator of one of the partial fractions."

"Are you sure you can factor every polynomial into linear factors?" the king said. "I thought . . ."

"Yes, you can factor anything," Recordis said. "Take my word for it. Except . . ." Recordis suddenly began to shiver, and then he trembled violently. "No—no—not those numbers!" he screamed, and he fell to the floor.

The king rushed to his assistance. "What happened to him?"

We were all puzzled, until the professor realized what had happened. "Recordis gets like this whenever he sees an *imaginary number,* such as $i = \sqrt{-1}$. You can factor any polynomial into linear factors, but some of the factors may contain terms that are not real numbers. For example, $(x^2 + 1) = (x - \sqrt{-1})(x + \sqrt{-1}) = (x - i)(x + i)$."

"We can't lose Recordis like this!" the king said. "We must find some way to avoid imaginary numbers."

"We could factor any polynomial into some linear factors and some quadratic factors," I suggested. "That way we will be assured that we won't have to deal with any imaginary numbers."

After repeated assurances that we would use only real numbers, Recordis finally calmed down, got up, and returned to normal. "This plan will create other problems, though," he said. "Suppose we have one linear factor and one quadratic factor in the denominator:

$$z = \frac{n_1 x^2 + n_2 x + n_3}{(a_1 x - b_1)(a_2 x^2 + b_2 x + c_2)}$$

"And you try partial fractions on that."

$$z = \frac{A}{a_1 x + b_1} + \frac{B}{a_2 x^2 + b_2 x + c_2}$$

$$n_1 x^2 + n_2 x + n_3 = A(a_2 x^2 + b_2 x + c_2) + B(a_1 x + b_1)$$
$$= Aa_2 x^2 + (Ab_2 + Ba_1)x + (Ac_2 + Bb_1)$$

coefficients of $x^2$:

$$n_1 = Aa_2$$

coefficients of $x$:

$$n_2 = Ab_2 + Ba_1$$

constant terms:

$$n_3 + Ac_2 + Bb_1$$

"Now we're in real trouble," Recordis said. "We have three equations but only two unknowns. There is no way we can find a solution."

We realized he had a good point. "There doesn't seem to be any way of getting rid of an equation," the professor said. "So it looks as though our only choice will be to find some place where we can add another variable to go along with $A$ and $B$." We experimented with a few possibilities until we were able to establish that, whenever there was a partial fraction with a quadratic denominator, the numerator would have to look like ($Ax + B$), with $A$ and $B$ both unknown.

"Let's see if it works," the professor said.

We tried an example:

$$y = \int \frac{3x^2 + 2x + 5}{(x - 1)(x^2 + 2x + 2)} \, dx$$

Setting up the partial fractions, we obtained:

$$\frac{3x^2 + 2x + 5}{(x - 1)(x^2 + 2x + 2)} = \frac{A}{x - 1} + \frac{Bx + C}{x^2 + 2x + 2}$$
$$= \frac{A(x^2 + 2x + 2) + (Bx + C)(x - 1)}{(x - 1)(x^2 + 2x + 2)}$$

Setting the numerators equal gave us:

$$3x^2 + 2x + 5 = Ax^2 + 2Ax + 2A + Bx^2 - Bx + Cx - C$$
$$= (A + B)x^2 + (2A - B + C)x + (2A - C)$$

coefficients of $x^2$:

$$3 = A + B$$

coefficients of $x$:

$$2 = 2A - B + C$$

constant terms:

$$5 = 2A - C$$

A bit of substitution made it possible to solve this three-equation, three-unknown system:

$$B = 3 - A$$
$$C = 2A - 5$$
$$\overline{\phantom{2 = 2A - (3 - A) + (2A - 5)}}$$
$$2 = 2A - (3 - A) + (2A - 5)$$
$$= 2A - 3 + A + 2A - 5$$
$$= 5A - 8$$
$$10 = 5A$$
$$A = 2$$
$$B = 1$$
$$C = -1$$

The result became:

$$\frac{3x^2 + 2x + 5}{(x - 1)(x^2 + 2x + 2)} = \frac{2}{x - 1} + \frac{x - 1}{x^2 + 2x + 2}$$

$$y = \int \frac{2}{x - 1}\, dx + \int \frac{x - 1}{x^2 + 2x + 2}\, dx$$

$$= 2\ln|x - 1| + \int \frac{x - 1}{x^2 + 2x + 2}\, dx$$

"I hate to disillusion you," Recordis said, "but we still don't know how to do that last integral."

It turned out that both Recordis and Trigonometeris were needed. Recordis realized that the substitution $u = x + 1$, $x = u - 1$, $du = dx$ could simplify the denominator by getting rid of the $2x$ term:

$$\int \frac{x-1}{x^2+2x+2}\,dx = \int \frac{u-2}{(u-1)^2+2(u-1)+2}\,du$$

$$= \int \frac{u-2}{u^2-2u+1+2u-2+2}\,du$$

$$= \int \frac{u-2}{u^2+1}\,du$$

Trigonometeris, in turn, realized that a trigonometric substitution would help for this integral: Let $u = \tan\theta$, $du = \sec^2\theta\,d\theta$.

$$\int \frac{x-1}{x^2+2x+2}\,dx = \int \frac{(\tan\theta-2)\sec^2\theta}{\tan^2\theta+1}\,d\theta$$

$$= \int \frac{(\tan\theta-2)\sec^2\theta}{\sec^2\theta}\,d\theta$$

$$= \int (\tan\theta-2)\,d\theta$$

$$\int \frac{x-1}{x^2+2x+2}\,dx = \ln|\sec\theta| - 2\theta + C$$

We had performed two substitutions in evaluating this integral: first we had substituted $u$ for $x$, and then we had substituted $\theta$ for $u$. That meant that we had to make two reverse substitutions to get our answer back in terms of $x$:

$$\int \frac{x-1}{x^2+2x+2}\,dx = \ln\sqrt{1+\tan^2\theta} - 2\theta$$

$$= \ln\sqrt{1+u^2} - 2\arctan u$$

$$= \tfrac{1}{2}\ln\left|[1+(x+1)^2]\right| - 2\arctan(x+1)$$

$$= \tfrac{1}{2}\ln\left|1+(x^2+2x+1)\right| - 2\arctan(x+1)$$

$$\int \frac{x-1}{x^2+2x+2}\,dx = \tfrac{1}{2}\ln\left|x^2+2x+2\right| - 2\arctan(x+1)$$

After doing all this work, we *still* had to write the final answer to the integral:

$$y = 2\ln|x-1| + \tfrac{1}{2}\ln\left|x^2+2x+2\right| - 2\arctan(x+1) + C$$

"I surely hope we never run into a denominator where we have to use a quadratic factor," Recordis said, gasping for breath. "Partial fractions was a nice method before we discovered that complication."

"I just thought of another case where we might not have enough unknowns," the king said. "Suppose that one of the factors in the denominator is repeated."

$$y = \frac{n_1 x^2 + n_2 x + n_3}{(x-a_1)(x-a_1)(x-a_2)}$$

"Then it won't work to set up the partial fractions as follows."

$$y = \frac{A}{x - a_1} + \frac{B}{x - a_1} + \frac{C}{x - a_2}$$

"$A$ and $B$ aren't really separate constants in this problem, because they both have the same denominator."

We realized that we had to make another adjustment in the partial fractions formula. It turned out that, whenever a factor in the denominator is repeated, one of the partial fractions has to have the *square* of that factor in the denominator:

$$\frac{n_1 x^2 + n_2 x + n_3}{(x - a_1)^2 (x - a_2)} = \frac{A}{x - a_1} + \frac{B}{(x - a_1)^2} + \frac{C}{x - a_2}$$

"We'd better make a summary of the method now, before anyone thinks of any more complications," Recordis said. (See p. 159.)

"That wraps up everything for today," Trigonometeris said, looking greatly relieved.

"You're not off the hook yet," Recordis told him. "You still need to solve your problem from yesterday. We started with this."

$$y = \int \frac{1}{\sqrt{a^2 + b^2 x^2}} \, dx$$

"And," he continued, "we ended up as follows."

$$x = \frac{a \tan \theta}{b}$$

$$y = \frac{1}{b} \int \sec \theta \, d\theta$$

"Let's use our result for the integral of the secant function," Trig said.

$$y = \frac{1}{b} \ln \left| \sec \theta + \tan \theta \right| + C$$

$$\tan \theta = \frac{bx}{a}$$

"We'll set up a triangle to find $\sec \theta$ (Figure 12–1)."

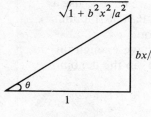

**Figure 12–1.**

$$\tan \theta = \frac{(\text{opposite side})}{(\text{adjacent side})} = \frac{bx}{a}$$

$$\sec \theta = \frac{(\text{hypotenuse})}{(\text{adjacent side})} = \left( 1 + \frac{b^2 x^2}{a^2} \right)^{1/2}$$

$$y = \int (a^2 + b^2 x^2)^{-1/2} \, dx$$

$$= \frac{1}{b} \ln \left| \left( 1 + \frac{b^2 x^2}{a^2} \right)^{+1/2} + \frac{bx}{a} \right| + C$$

## EVALUATING INTEGRALS BY PARTIAL FRACTIONS

The method of partial fractions is useful when the integrand contains a rational function, that is, a fraction with a polynomial in the numerator of degree less than the degree of the polynomial in the denominator:

$$\int \frac{a_m x^m + a_{m-1} x^{m-1} + a_{m-2} x^{m-2} + \cdots + a_1 x + a_0}{b_n x^n + b_{n-1} x^{n-1} + b_{n-2} x^{y-2} + \cdots + b_1 x + b_0}\, dx \quad (m < n)$$

The goal of the method is to break the integrand into a sum of fractions that are much simpler. The first step is to factor the denominator. The result will be a product of some linear factors and some quadratic factors. All the numbers that result will be real, but there is no guarantee that they will be rational. (Of course, if the denominator comes to you already factored, you will be saved a *lot* of work.) The integrand can then be resolved as a sum of partial fractions as follows:

1. If a linear factor (such as $ax + \cdot b$) occurs once in the denominator, then there will be a partial fraction of the form $A/(ax + b)$.
2. If the linear factor $(ax + b)$ occurs $k$ times in the denominator, then there are $k$ partial fractions of the form $A_1/(ax + b)$, $A_2/(ax + b)^2$, . . . , $A_k/(ax + b)^k$.
3. If a quadratic factor (such as $ax^2 + bx + c$) occurs once in the denominator, then there is a partial fraction of the form $(Ax + B)/(ax^2 + bx + c)$. (Note that $(b^2 - 4ac)$ is negative in this case. Otherwise, the quadratic factor can be broken into a product of two linear factors.)
4. If the quadratic factor $(ax^2 + bx + c)$ occurs $j$ times in the denominator, then there are $j$ partial fractions of the form $(A_1 x + B_1)/(ax^2 + bx + c)$,

   $(A_2 x + B_2)/(ax^2 + bx + c)^2$, . . . $(A_j x + B_j)/(ax^2 + bx + c)^j$.

The numerators of the partial fractions must be solved for next. Once the integrand has been broken up into partial fractions, each integral can be solved individually. The integrals with linear denominators can be solved with logarithms, and the integrals with quadratic denominators can be solved with trigonometric substitution, using the secant-tangent identity.

All of us, especially Trigonometeris, were greatly relieved when we finally finished the work for that day. One result was that Recordis and Trigonometeris realized they were very dependent on each other. After that day Recordis was less likely to make snide remarks about trigonometry, and Trigonometeris was less likely to belittle algebra.

## Note to Chapter 12

The *degree* of a polynomial is the highest power that appears in the polynomial. For example, $(x + 1)$ is a first-degree polynomial, $(x^2 + 2x + 3)$ is a second-degree polynomial, and $(x^3 + 3x^2 + 4)$ is a third-degree polynomial. A *rational function* is a fraction in which both the numerator and denominator are polynomials. If the degree of the numerator is less than the degree of the denominator, the rational function is called a *proper rational function*. If the degree of the numerator is greater than the degree of the denominator, the function is called an *improper rational function*. An improper rational function can always be written as the sum of a polynomial plus a proper rational function. Note the analogy with real numbers. A fraction $n/d$ is called a *proper fraction* if $n < d$, and an *improper fraction* if $n > d$. An improper fraction can always be written as the sum of an integer and a proper fraction. (For example, $\frac{3}{2}$ can be written as $1 + \frac{1}{2}$.)

## Exercises

Express each of these fractions as a sum of partial fractions:

**1.** $\dfrac{x + 4}{(x - 3)(x - 2)}$       **2.** $\dfrac{3x}{(x + \frac{1}{2})(x - \frac{1}{2})}$

**3.** $\dfrac{2x - 4}{(2x - 2)(4x + 3)}$       **4.** $\dfrac{1}{x(x + 3)}$

**5.** $\dfrac{x^2 + x + 1}{x^3 + 3x^2 + 3x + 1}$

Combine into single fractions:

**6.** $\dfrac{7.1}{10 - 0.6z} - \dfrac{6.4}{9 - 0.4z}$       **7.** $\dfrac{1}{2 - 2x} + \dfrac{1}{2 + 2x}$

**8.** $\dfrac{2}{x - 1} + \dfrac{x - 1}{x^2 + 2x + 2}$

Solve for $y$:

**9.** $\dfrac{dy}{dx} = \dfrac{2x - 7}{(x - 4)(x - 3)}$       **10.** $\dfrac{dy}{dx} = \dfrac{2x + 9}{(\frac{1}{2}x + 5)(x - 1)}$

**11.** $\dfrac{dy}{dx} = \dfrac{(1 - \sqrt{3})x - 4}{(x - \sqrt{3})(x - 1)}$       **12.** $\dfrac{dy}{dx} = \dfrac{x}{x^2 - 1}$

Differentiate each function to verify an integration result from the chapter:

**13.** $y = \frac{1}{2} \ln\left(\dfrac{1 + x}{1 - x}\right)$

**14.** $y = \ln(\sec x + \tan x)$

**15.** $y = -\ln \cos x$

**16.** $y = \frac{1}{2} \ln\left|(x^2 + 2x + 2)\right| - 2 \arctan(x + 1) + 2 \ln\left|x - 1\right|$

**17.** Evaluate: $y = \int \sec^3 x\ dx$. Verify the result by differentiation.

**18.** Find the area of the section of the hyperbola $x^2/a^2 - y^2/b^2 = 1$ that is bounded by the curve and the line $x = 2a$. (See Figure 12–2.)

Evaluate:

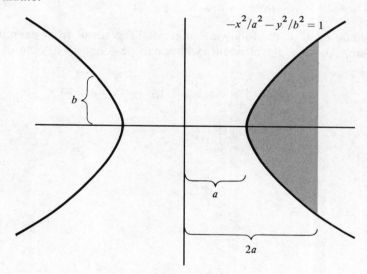

$$-x^2/a^2 - y^2/b^2 = 1$$

**Figure 12–2.**

**19.** $y = \int \sqrt{1 + x^2}\ dx$. (Use the result from exercise 17.)

**20.** $y = \int (1 + x^2)^{-1/2}\ dx$

**21.** $y = \displaystyle\int \frac{\sqrt{1 - x^2}}{x}\,dx$

**22.** $y = \displaystyle\int \frac{1}{x\sqrt{1 + x^2}}\ dx$

**23.** $y = \displaystyle\int \frac{x^2}{\sqrt{1 + x^2}}\ dx$

Express each improper rational function as the sum of a polynomial and a proper rational function:

**24.** $\dfrac{2x^4 + 5x^3 + 7x^2 + 5x}{2x^2 + 3x + 4}$

**25.** $\dfrac{15x^4 - 15x + 1}{3x^3 - 3}$

**26.** $\dfrac{5x^4 + 4x^3 - 5x^2 - 4x + 1}{x^2 - 1}$

**27.** $\dfrac{3x^4 - x^3 - 33x^2 - 25x - 61}{x^2 - x - 12}$

**28.** Find a general formula for the integral

$$y = \int \frac{1}{ax^2 + bx + c}\, dx$$

Consider two cases: (a) $b^2 - 4ac > 0$. Define a new constant $D$ so that $D = \sqrt{b^2 - 4ac}$. (b) $b^2 - 4ac < 0$. Let $D = \sqrt{4ac - b^2}$.

Evaluate each of the following integrals. The result from exercise 28 will help. Another useful result, which can be found in a table of integrals, is:

$$\int \frac{Ax + B}{ax^2 + bx + c}\, dx = \frac{A}{2a} \ln\left| ax^2 + bx + c \right|$$
$$+ \left( \frac{B - Ab}{2a} \right) \int \frac{1}{ax^2 + bx + c}\, dx$$

**29.** $\displaystyle\int \frac{1}{x^2 + 2x + 2}\, dx$

**30.** $\displaystyle\int \frac{1}{x^2 - 3x - 1}\, dx$

**31.** $\displaystyle\int \frac{1}{x^2 + x + 1}\, dx$

**32.** $y = \displaystyle\int \frac{1}{x^2 - x + 1}\, dx$

**33.** $y = \displaystyle\int \frac{1}{x^3 - 1}\, dx$

**34.** $y = \displaystyle\int \frac{1}{x^3 + 1}\, dx$

**35.** $y = \displaystyle\int \frac{1}{2x^2 + 2x}\, dx$

**36.** $y = \displaystyle\int \frac{1}{3x^2 + 5x + 1}\, dx$

**37.** $y = \displaystyle\int \frac{x}{x^4 - 1}\, dx$

**38.** $y = \displaystyle\int \frac{1}{(x - 1)(x^2 + 1)}\, dx$

# 13
# Finding Volumes with Integrals

Plans for the party were proceeding smoothly, but there were still a lot of arrangements to be made. Every day Recordis or Builder came up with a new problem to be solved. "There are so many little things that need doing," Recordis kept saying.

The professor was working on her outline of calculus. She was convinced that we were completely finished with the subject. The king, though, was becoming uneasy. "I'm sure there's something we're missing," he said. "Definite integrals must be good for more than just finding area."

"No way!" the professor said. "We know that integrals are good for two things—finding antiderivatives and finding areas."

At that moment Builder came into the room in a very frustrated state of mind. "I'm worried about Column Mountain," he said. "That's the mountain with the perfect paraboloid shape at its top." (See Figure 13–1.)

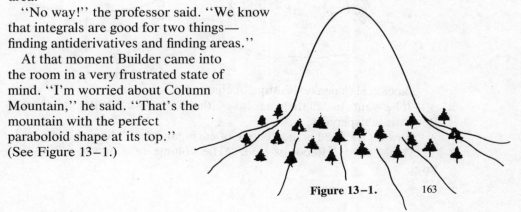

**Figure 13–1.**          163

"I remember what a paraboloid is," the professor said. "You make one by taking a regular two-dimensional parabola and rotating it in three dimensions."

"The top of the mountain is very unstable," Builder told us, "and we're afraid that the children at the party will jump around so much that the mountain could collapse. So we've decided that we'll remove the dirt from the top of the mountain. Before we start, I need to know the volume of the dirt we have to remove."

"It's too bad we can't help you," Recordis said. "Of course, if you needed to know the area of something . . ."

"We should be able to do something," the king protested, and he started to call the others.

"No, we can't do anything," Recordis said. "I just flipped through all my old tables. We don't have any formula for the volume of a paraboloid."

Builder gave us the exact measurements of the mountain anyway. The cross section of the mountain obeyed the relationship $y = -\frac{1}{2}x^2$, and we needed to remove the dirt from $y = 0$ to $y = -100$.

"I know how we can approximate the volume," the king said. "Let's start in the same way that we started to find the area under a curve. (See Chapter 8.) We can divide the mountain up into a series of little cylinders." (Figure 13–2.)

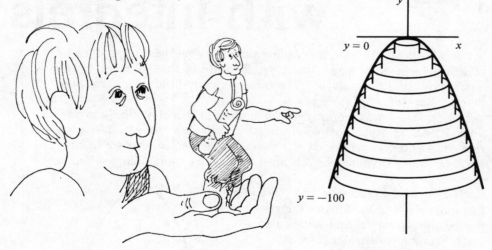

**Figure 13–2.**

"Pancakes! Pancakes!" Mongol chanted.

("If he wants to call them pancakes, that's what we'd better call them," Recordis whispered.)

"Now all we need is the volume of each pancake," the professor said.

"That's easy," Recordis said. "The volume of a pancake is as follows."

$$\text{(base area)} \times \text{(height)} = \pi(\text{radius})^2(\text{height})$$

"In this case $x$ is the radius," the professor said.

$$\text{(volume of pancake)} = \pi x^2(\text{height})$$

"What are we going to call the height?" Recordis asked.

"What are we going to call the volume?" the king asked.

"We could call the height $\Delta y$," I suggested. "And we could call the volume of each little pancake $\Delta v_i$, where the little $i$ stands for the $i$th pancake."

$$\Delta v_i = \pi x^2 \, \Delta y$$

"Then to get the approximate volume of the paraboloid we just add up the volumes of all the pancakes," the professor said.

$$\text{(volume of paraboloid)} \simeq \sum_{i=1}^{n} \Delta v_i = \sum_{i=1}^{n} \pi x^2 \, \Delta y$$

"We're still missing a lot of volume," Recordis objected. "Look at how much volume is part of the paraboloid that isn't part of any of the pancakes."

"Then we can add more and more pancakes, so the approximate volume comes closer to the total volume," the professor said.

Recordis was about to protest that he would have to add together all of the pancakes, but before he could the king cried out, "I know how we can get the exact volume of the paraboloid! We take the limit as the number of pancakes goes to infinity!"

$$\text{(volume of paraboloid)} = \lim_{n \to \infty, \, \Delta y \to 0} \sum_{i=1}^{n} \pi x^2 \, \Delta y$$

"We could call that the *continuous sum*," the professor said. "Before we take the limit we can say that we have a *discrete sum*."

We were all impressed that we had proceeded this far, but we realized that we did not know where to go next.

"This won't work!" Recordis said emphatically. "In fact, we have already proved that this doesn't work! This is exactly like the first method we tried to find the area of my pools, and we know that didn't work at all!"

"That's it!" the professor exclaimed suddenly.

"What's it?" Recordis said.

"What you just said!" the professor went on excitedly. "We already found that the area under a curve is equal to the continuous sum."

$$\text{(area under curve)} = \lim_{\Delta x \to 0} \sum_{i=1}^{n} f(x_i) \, \Delta x$$

"I know that!" Recordis said. "But we don't know how to calculate a continuous sum."

"But we also found that the area under a curve is equal to the definite integral."

$$(\text{area under curve}) = \int_a^b f(x)\ dx$$

"I already know that," Recordis said.

"If both the continuous sum and the definite integral equal the area, then they must be equal to each other!" the professor proclaimed triumphantly.

$$\lim_{\Delta x \to 0} \sum_{i=1}^{n} f(x_i)\ \Delta x = (\text{continuous sum}) = (\text{area})$$

$$= (\text{definite integral}) = \int_a^b f(x)\ dx$$

$$= \int_a^b f(x)\ dx$$

"Now we can evaluate the continuous sum," the king said. "We just need to solve the definite integral, and we know how to do that. And the continuous sum can represent anything, instead of just an area. It can represent a volume or even something else."

"There is an uncanny similarity between the two notations," the professor said. "Notice how this part, the crooked $s$, $\Sigma$, corresponds to the curvy part, which is almost like a letter $s$: $\int$. Notice also how the $f(x_i)$ corresponds to the $f(x)$ part, and the $\Delta x$ part corresponds to the $dx$ part."

$$\lim_{\Delta x \to 0} \sum_{i=1}^{n} f(x_i)\ \Delta x$$

$$\int_a^b f(x)\ dx$$

"That's amazing," the king said. "I wonder how that happened?"

"It almost looks as though the integral sign was set up to stand for a continuous sum," Recordis remarked, looking at me suspiciously.

"Can you solve the problem now?" Builder asked.

We wrote our expression for the volume of the paraboloid as a continuous sum:

$$V = \lim_{\Delta y \to 0} \sum_{i=1}^{n} \pi x^2\ \Delta y$$

"Now it's easy to turn this into a definite integral," the king said.

$$V = \int \pi x^2\ dy$$

"What about the limits of integration?" Recordis asked.

"The limits must be in terms of $y$, since we are integrating along $y$," the king said. "We need to set up the limits so that our pancakes will cover all of the volume that we want."

"It looks as though we start integrating where $y = -100$ and keep integrating until $y = 0$," the professor said. (See Figure 13–2.)

$$V = \pi \int_{-100}^{0} x^2 \, dy$$

"Now we're stuck," Recordis said. "We have the integrand written in terms of $x$, but the $d$-variable term is in terms of $y$."

Builder told him, "I already told you that $y = -\frac{1}{2}x^2$."

We put that expression into the integral, and then we found everything was quite straightforward:

$$x^2 = -2y$$
$$V = \pi \int_{-100}^{0} (-2y) \, dy$$
$$= (-2\pi)(\tfrac{1}{2}y^2) \Big|_{-100}^{0}$$
$$= 10{,}000\pi$$

"That's a lot of dirt," Builder whistled. "I'll take your word for it."

"I don't take your word for it," Recordis disagreed. "I'm not sure that this pancake method really works. Let's try it for something that we know the volume of already, such as a sphere of radius $r$."

"We know that the volume of a sphere is $(4/3)\pi r^3$," the king said.

The professor started to divide the sphere into pancakes. "Let's find the volume of one hemisphere first," she suggested. "Then we can multiply by 2 to get the volume of the whole sphere." (Figure 13–3.)

**Figure 13–3.**

I suggested that we call the volume of a single pancake $dV$, to make it easier to set up the integral $\int dV = V$. We decided to call the height of each pancake $dy$, because we would be integrating along $y$. Then the volume of each pancake would be $dV = \pi x^2 \, dy$. We set up the integral:

$$V = \int dV = \int \pi x^2 \, dy$$

We decided that the limits of integration, which had to be written in terms of $y$, would be $y = 0$ to $y = r$:

$$V = \int_0^r \pi x^2 \, dy$$

Now we had to solve for $x^2$ in terms of $y$. Recordis recognized an algebra problem when he saw it. Since the cross section of a sphere is a circle, he used the equation of a circle:

$$x^2 + y^2 = r^2$$
$$x^2 = r^2 - y^2$$
$$V = \pi \int_0^r (r^2 - y^2) \, dy$$

Now it was just a matter of using the methods we had developed to solve definite integrals:

$$V = \pi \int_0^r r^2 \, dy - \pi \int_0^r y^2 \, dy$$
$$= \pi r^2 y \Big|_0^r - \pi(\tfrac{1}{3}) y^3 \Big|_0^r$$
$$= \pi r^2 (r - 0) - \pi(\tfrac{1}{3})(r^3 - 0^3)$$
$$= \pi r^3 - \tfrac{1}{3}\pi r^3$$
$$V = \tfrac{2}{3}\pi r^3$$

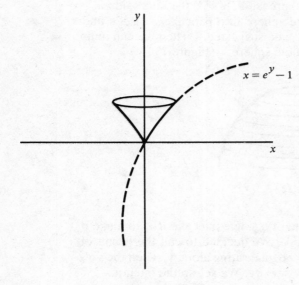

Figure 13–4.

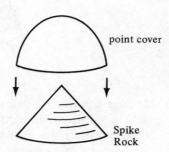

Figure 13–5.

"Just what we wanted!" the professor said with relief. "If the volume of a hemisphere is $\frac{2}{3}\pi r^3$, then the volume of a whole sphere is $\frac{4}{3}\pi r^3$."

"So this method does work!" Recordis rejoiced. "I have a whole bunch of party arrangements that I'm working on that this method could help with." He flipped to a page in one of his notebooks.

"Here is a good example. We designed special ice cream cones to give to the children at the party. If we could figure out the volume of the cones, then we would know how much ice cream we will need to get."

The curved ice cream cones were formed by rotating the curve $x = e^y - 1$ about the $y$ axis, from $y = 0$ to $y = h$ (Figure 13–4). (We used $h$ to stand for the height of the ice cream cones because Recordis had forgotten what the height was.)

"We can use pancakes easily," the professor said. "We will be integrating along $y$, from $y = 0$ to $y = h$."

The volume of each pancake was given by:

$$dV = \pi x^2 \, dy$$

$$x = e^y - 1$$

$$dV = \pi (e^y - 1)^2 \, dy$$

$$V = \pi \int_0^h (e^{2y} - 2e^y + 1) \, dy$$

$$= \pi \left( \tfrac{1}{2} e^{2y} - 2e^y + y \right) \Big|_{y=0}^{h}$$

$$= \pi ( \tfrac{1}{2} e^{2h} - 2e^h + h - \tfrac{1}{2} e^0 + 2e^0 - 0 )$$

$$V = \pi ( \tfrac{1}{2} e^{2h} - 2e^h + h + \tfrac{3}{2} )$$

"I have another problem we should be able to solve now," the king said. "The gardener mentioned it to me. Remember Spike Rock, the rock with the perfect conical top next to the rose garden (Figure 13–5)? I am very much afraid that one of the children at the party could be hurt unless we do something to cover up the spike. So Builder and I figured out how to build a round point-cover out of clay to put on the spike. If we can figure out the volume of the point-cover, we will know how much clay we need to buy."

It turned out that the point-cover was formed by taking the region bounded by the two curves $y = x$ and $y = x^2$ and rotating that region about the $y$ axis (Figure 13–6). (Notice that we turned the point-cover upside down to make it a bit easier to figure out the volume.)

"I hate to disillusion you," Recordis said, "but there is no way that we can fit any pancakes into a shape like that."

"Do we have to use pancakes?" the king asked. "Maybe we could use another shape."

"How?" Recordis inquired.

"All we need to do is set up the continuous sum," the professor agreed. "We might be able to find another shape that works."

I suggested that we try using little cylindrical shells (Figure 13–7). "If the shell is thin enough, we can say that the volume of the shell is approximately equal to $dV = (2\pi r)(h)(dx)$." (Note that $2\pi r$ is the circumference of the cylinder, so $2\pi rh$ is the outer surface area of the cylinder. The volume of the shell is then equal to the surface area times the thickness.)

"You had better make sure that you don't let the shells become too thick," Recordis said. "If the shells are too thick, the approximate formula you are using for the volume won't work."

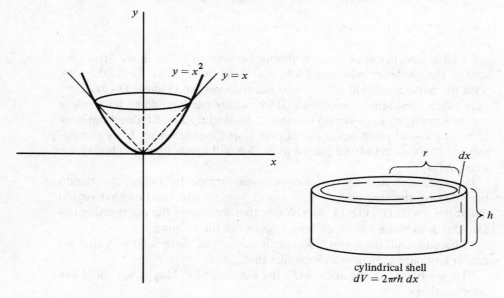

**Figure 13–6.**

cylindrical shell
$dV = 2\pi rh\,dx$

**Figure 13–7.**

"That shouldn't be any problem," the professor told him. "Remember that we are going to take the limit where the thickness of the shells goes to zero."

Next, we had to fit a representative cylindrical shell into the point-cover (Figure 13–8).

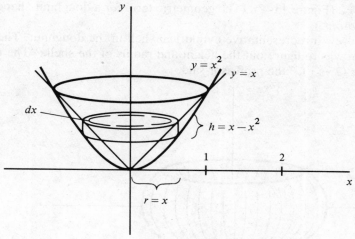

**Figure 13–8.**

"Now we need the radius of the shell and the height of the shell," the professor said. The radius turned out to be simply equal to the $x$ coordinate of the shell, and the height of the shell was equal to the distance between the two curves: $h = x - x^2$.

"Now we have to put all this together into a sum," the king stated.

$$V = \lim_{\Delta x \to 0} \sum_{i=1}^{n} 2\pi r h \; \Delta x$$

$$= \lim_{\Delta x \to 0} 2\pi \sum_{i=1}^{n} x(x - x^2) \; \Delta x$$

$$= 2\pi \int_{0}^{1} x(x - x^2) \; dx$$

"That's an easy integral," Recordis said.

$$V = 2\pi \int_{0}^{1} (x^2 - x^3) \; dx$$

$$= 2\pi \left. \tfrac{1}{3}x^3 \right|_{0}^{1} - \left. \tfrac{1}{4}x^4 \right|_{0}^{1}$$

$$= 2\pi(\tfrac{1}{3} - \tfrac{1}{4})$$

$$V = \frac{\pi}{6}$$

"It's a good thing that we just invented the method of cylindrical shells," Reçordis said. "It will help with something else I need to do for the party. We're giving out doughnuts to the children. I have been trying to figure out what the volume of a doughnut is, so that I will know how much doughnut mix I need to order."

The doughnuts were formed by rotating the circle $x^2 + y^2 = 1$ about the line $x = 2$ (Figure 13–9). (The geometric term for a doughnut-shaped figure is *toroid*.)

Igor drew a representative cylindrical shell in the doughnut. The next problem was to figure out the height and radius of the shell. "The radius must be $(2 - x)$," the professor said.

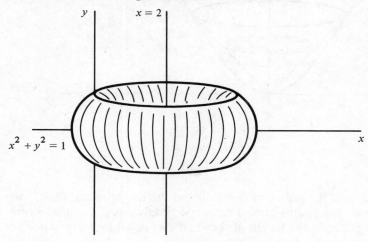

**Figure 13–9.**

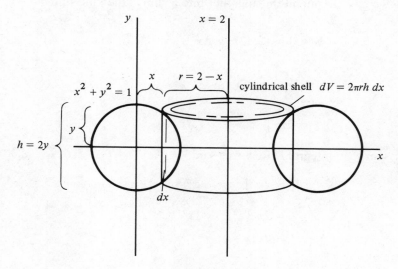

**Figure 13–10.**

"The height must be $2y$," the king added (Figure 13–10).
We set up the equation for the volume of the cylindrical shell:

$$dV = 2\pi rh$$

$$r = 2 - x$$

$$h = 2y$$

"We can solve for $y$," Recordis noted.

$$x^2 + y^2 = 1$$

$$y^2 = 1 - x^2$$

$$y = \sqrt{1 - x^2}$$

$$dV = 2\pi(2 - x)\, 2\sqrt{1 - x^2}\, dx$$

"Now we need to figure out the limits of integration," the professor said. "They must be in terms of $x$, since we are integrating along $x$." After looking closely at the diagram, we decided that we could catch all of the volume of the doughnut with our cylindrical shells if we integrated from $x = -1$ to $x = 1$.

$$V = \int_{-1}^{1} 2\pi(2 - x)\, 2\sqrt{1 - x^2}\, dx$$

$$= 4\pi \int_{-1}^{1} (2 - x)\, \sqrt{1 - x^2}\, dx$$

$$= 4\pi \int_{-1}^{1} 2\sqrt{1 - x^2}\, dx - 4\pi \int_{-1}^{1} x\sqrt{1 - x^2}\, dx$$

We made the substitution $u = 1 - x^2$, $du = -2x\, dx$ for the second integral, with the result:

$$V = 8\pi \int_{-1}^{1} \sqrt{1 - x^2}\, dx + 2\pi \int_{0}^{0} u^{1/2}\, du$$

"That second integral can't be right!" Recordis exclaimed. "That's equal to zero!"
    After we had looked at the integral for a while, the professor said, "Of course it must be zero. Look at a graph of the integrand (Figure 13–11). Half of the graph is above the $x$ axis, so its area is positive; but exactly half of the graph is below the $x$ axis, so its area is negative. The two areas cancel out, so the total value of the definite integral must be zero."
    "Anyway, we know that the volume of the doughnut can't be zero," Recordis said.

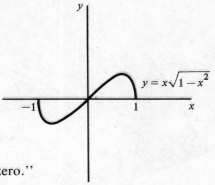

Figure 13–11.

"It's lucky for us that we still have the first integral to evaluate."

$$V = 8\pi \int_{-1}^{1} (1 - x^2)^{1/2} \, dx$$

We made the trigonometric substitution $x = \sin\theta$, $dx = \cos\theta \, d\theta$.

$$V = 8\pi \int_{-\pi/2}^{\pi/2} \cos^2\theta \, d\theta$$

We had done the integral of $\cos^2\theta$ before (see Chapter 11), so the final answer for the volume turned out to be:

$$V = 8\pi\left(\frac{\pi}{2}\right)$$

$$= 4\pi^2$$

"I have another problem related to the party," Builder said. "We want to build a grandstand in one corner of the auditorium (Figure 13–12). The height of the grandstand is $h$. The base of the grandstand is an isosceles right triangle, with each side equal to $s$. I'm sure you can figure out the volume under the grandstand."

"We can?" Recordis asked. "That's not a figure of revolution, so it's not like anything else we've done. There is no way that you could fit either pancakes or cylindrical shells into a pyramid like that. In fact, the only shape that you could fit in there would be little triangles."

"That's a good idea," the professor said. "We'll use little triangles." (Figure 13–13.)

The volume of each little triangular segment was easy to figure out, since it was just the area of the triangle times the thickness. "With a right triangle it's easy to find the area," Recordis noted. "You just multiply together the two legs and divide by 2." In our case the two legs were equal, so we just needed to find one of the legs. We called $x$ the length of one leg for our representative triangle, and we called $z$ the distance down from the top of the pyramid-shaped grandstand (Figure 13–14).

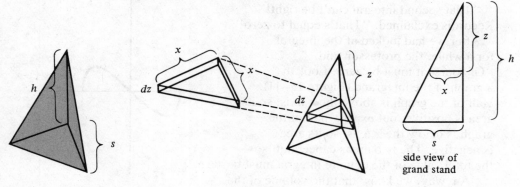

Figure 13–12.          Figure 13–13.          Figure 13–14.

"I know how we can find $x$," the professor said. "We can use a similar triangle relationship to establish that $z/h = x/s$."

$$x = \frac{zs}{h}$$

We wrote down the volume of the differential triangular element:

$$dV = \tfrac{1}{2}x^2 \, dz = \frac{s^2}{2h^2} z^2 \, dz$$

We integrated along $z$, from $z = 0$ to $z = h$:

$$V = \int_0^h \frac{s^2 z^2}{2h^2} \, dz$$

$$= \frac{s^2}{2h^2} \int_0^h z^2 \, dz$$

$$= \frac{s^2}{2h^2} \left. \tfrac{1}{3}z^3 \right|_0^h$$

$$= \frac{s^2 h^3}{6h^2}$$

$$= \frac{s^2 h}{6}$$

"I never thought an integral could be so versatile," Recordis said. "This will be the best party ever, and we've made calculus do all the work. Now I had some other problems that I think we might be able to solve if I could only remember what they were . . ."

## Exercises

1. Find the volume of the solid formed by rotating the ellipse $x^2/a^2 + y^2/b^2 = 1$ about the $x$ axis.
2. The base of a solid is the region bounded by the parabola $y = \frac{1}{2}x^2$ and the line $y = 2$. Each plane section of the solid perpendicular to the $y$ axis is an equilateral triangle. Find the volume of the solid.
3. A paraboloid dish (cross section $y = x^2$) is 8 units deep. It is filled with water up to a height of 4 units. How much water must be added to the dish to fill it completely?
4. Write the integral that represents the volume of the solid formed by rotating the region bounded by $y = f(x)$, $x = a$, $x = b$, and $y = 0$ about (a) the $x$ axis; (b) the line $x = b$; and (c) the line $x = c$, where $c > a, b$.
5. Consider the solid formed by rotating the curve $y = f(x)$ from $x = a$ to $x = b$ about the $x$ axis. Let $V(x)$ be the function whose value is the volume of the solid between $x = a$ and $x = x$. (a) What is $V(a)$? (b) For some small $\Delta x$, what is $V(x + \Delta x) - V(x)$? (c) Using the definition of the derivative, find the definite integral that represents the total volume of the solid. (Let $F(x)$ be a function such that $dF/dx = \pi[f(x)]^2$.)
6. Find the volume of the top quarter of a sphere:

$$V = \int_{r/2}^{r} \pi x^2 \, dy$$

7. Find the volume of the sphere by integrating from $y = -r$ to $y = r$. Compare the result with the one arrived at by taking twice the integral from $y = 0$ to $y = r$.
8. Find the volume of a pyramid with square base and sides that are equilateral triangles.

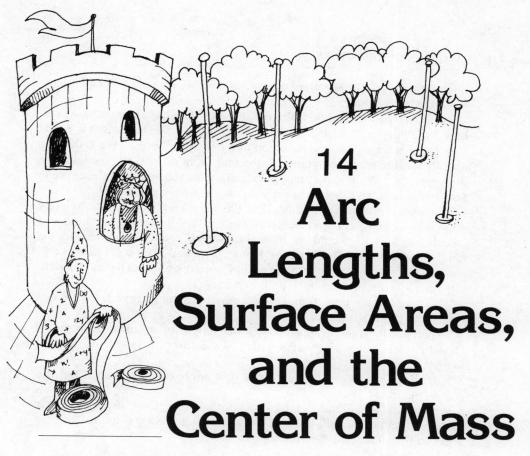

# 14
# Arc Lengths, Surface Areas, and the Center of Mass

The next day we were in the courtyard preparing to cut ribbons to use as streamers around the party site. Builder had already built poles throughout the grounds from which he could hang the ribbons. The professor unrolled the ribbons while Recordis stood next to the roll, prepared to cut them.

"How long should we make the ribbons?" Recordis called out to the professor.

The professor thought for a moment. "We had better not cut them too long, or else they will hang down too far and will get in the way when the adults walk under them."

"But we can't cut them too short," Recordis said. "We want the ribbons to hang down a little bit, so they make a nice curve."

"We should be able to figure out the exact length," the king told them. "Builder, how far apart are the poles?"

"They're all 4 meters apart, and each pole is 2.2 meters tall," Builder answered.

"The lowest part of the hanging ribbon should be 2 meters high," Trigonometeris said. "That will be high enough so that all the adult heads will fit under it."

"We don't know the equation of a hanging ribbon, and we don't know how to figure out the length of a curve even if we do know its equation,'' Recordis pointed out.

"Builder, you should be able to figure out the shape of a hanging ribbon,'' the king said. Builder nodded hesitantly, as if he knew that it would be a lot of work to find the equation of a hanging ribbon. "We should be able to find the length of a curve,'' the king went on. "We'll use what we did yesterday. We'll make an approximation for the length of the curve in terms of a sum, and then turn the sum into a definite integral.''

The professor was irritated because this meant we were not completely done with calculus yet, and Recordis was irritated because it sounded as if calculating lengths would be hard, but we all prudently agreed to the king's plan anyway. Builder took some string and returned to his workroom. The rest of us went into the Main Conference Room to set up a definite integral to represent the length of a curve.

"We don't know how to find the length of anything curved, except circles,'' the professor said.

"Then we'll have to approximate the length of a curve in terms of straight lines,'' the king suggested. "We know the length of straight lines.''

Igor drew a series of straight lines that approximately traced out the curve (Figure 14–1).

We decided to call the length of each little line segment $ds$. It was clear that the length of the whole curve (which we called $S$) was approximately equal to the sum of all the little lengths $ds$:

$$S \simeq \sum_{i=1}^{n} ds_i$$

"We can get the exact length by taking the limit as $ds$ goes to zero,'' the professor noted.

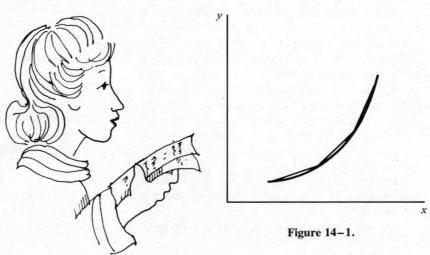

**Figure 14–1.**

$$S = \lim_{ds \to 0} \sum_{i=1}^{n} ds_i$$

"And that, of course, is equal to the definite integral."

$$S = \int ds$$

"We can't do very much with that integral when it is written in terms of $s$," Recordis said. "For one thing, we don't know what the limits of integration are in terms of $s$. We need to express the integral in terms of $x$."

Igor displayed a close-up version of one of the little segments $ds$ (Figure 14–2).

"We can call the distance that $ds$ goes up $dy$, and the distance that it goes sideways $dx$," the professor said. "That makes use of differential notation."

"We can use the Pythagorean theorem to express $ds$ in terms of $dx$ and $dy$," the king suggested.

$$ds^2 = dx^2 + dy^2$$
$$ds = \sqrt{dx^2 + dy^2}$$

"Do we have to bring the Pythagorean theorem into this?" Recordis complained. "Whenever you use that theorem, you end up with a square root sign. Wouldn't it be easier to just make the approximation that, as $ds$ goes to zero, $ds$ and $dx$ are approximately equal? Then we could just integrate $dx$ to get the length of the curve." Recordis' method certainly sounded simpler, until we realized that it would give the same result for the lengths of all the curves in Figure 14–3. Sadly we went back to the method that had the square root sign.

$$(\text{length}) = S = \int ds = \int \sqrt{dx^2 + dy^2}$$

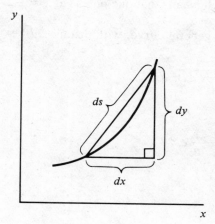

**Figure 14–2.**

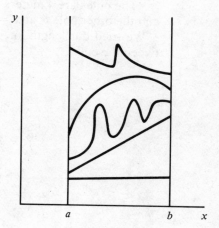

**Figure 14–3.**

The professor suggested a way to modify the square root so that we could integrate along *x:*

$$S = \int \sqrt{(dx^2)[1 + (dy/dx)^2]}$$
$$= \int \sqrt{1 + (dy/dx)^2} \; dx$$

"We'd better make sure that we remember the difference between the second derivative $(d^2y/dx^2)$ and the square of the first derivative $(dy/dx)^2$," Recordis cautioned.

"That's the integral we need!" the king said. "Now all we have to do is write the limits in terms of *x.*"

$$(\text{length of curve from } x = a \text{ to } x = b) = S = \int_a^b \sqrt{1 + (dy/dx)^2} \; dx$$

"With this integral we can find the length of any curve, as long as we know *dy/dx.*"

Just then Builder came into the room with the answer to the hanging string problem. "This curve turns out to be a weird one, but I guarantee this is the right curve. The general shape of a hanging ribbon is given by this expression."

$$y = \tfrac{1}{2}a(e^{x/a} + e^{-x/a})$$

(This curve is known as a *catenary.* It can also be expressed using a function known as the *hyperbolic cosine,* defined by $\cosh x = \tfrac{1}{2}(e^x + e^{-x})$.)

"That number *e* again!" Recordis said. "I am always amazed how many times that one number turns up."

"In your particular case $a = 10$," Builder told him. "Your specific curve is given by:

$$y = 5(e^{x/10} + e^{-x/10}) - 8$$

"The poles are 4 meters apart, which means that one pole is at $x = -2$ and the other pole is at $x = 2$."

We tried our length integral on the general curve, with distance *d* between the ends:

$$S = \int_{-d/2}^{d/2} \sqrt{1 + (dy/dx)^2} \; dx$$

$$y = \tfrac{1}{2}a(e^{x/a} + e^{-x/a})$$

$$\frac{dy}{dx} = \tfrac{1}{2}a\left(\frac{1}{a} e^{x/a} - \frac{1}{a} e^{-x/a}\right)$$

$$= \tfrac{1}{2}(e^{x/a} - e^{-x/a})$$

$$\left(\frac{dy}{dx}\right)^2 = \tfrac{1}{4}(e^{2x/a} - 2e^{x/a}e^{-x/a} + e^{-2x/a})$$

$$= \tfrac{1}{4}(e^{2x/a} - 2 + e^{-2x/a})$$

$$S = \int_{-d/2}^{d/2} \sqrt{1 + \tfrac{1}{4}(e^{2x/a} - 2 + e^{-2x/a})} \, dx$$

$$= \int_{-d/2}^{d/2} \sqrt{1 + \tfrac{1}{4}e^{2x/a} - \tfrac{1}{2} + \tfrac{1}{4}e^{-2x/a}} \, dx$$

$$= \int_{-d/2}^{d/2} \sqrt{\tfrac{1}{4}e^{2x/a} + \tfrac{1}{2} + \tfrac{1}{4}e^{-2x/a}} \, dx$$

"If only we could get rid of the square root sign!" Recordis said. "We can," the professor assured him. "Watch this."

$$S = \int_{-d/2}^{d/2} \tfrac{1}{2}\sqrt{e^{2x/a} + 2 + e^{-2x/a}} \, dx$$

$$(e^{x/a} + e^{-x/a})^2 = e^{2x/a} + 2 + e^{-2x/a}$$

$$S = \int_{-d/2}^{d/2} \tfrac{1}{2}\sqrt{(e^{x/a} + e^{-x/a})^2} \, dx$$

$$= \tfrac{1}{2} \int_{-d/2}^{d/2} (e^{x/a} + e^{-x/a}) \, dx$$

$$= \tfrac{1}{2}ae^{x/a} \Big|_{-d/2}^{d/2} - \tfrac{1}{2}ae^{-x/a} \Big|_{-d/2}^{d/2}$$

$$= \tfrac{1}{2}ae^{d/2a} - \tfrac{1}{2}ae^{-d/2a} + \tfrac{1}{2}ae^{-d/2a} - \tfrac{1}{2}ae^{-d/2a}$$

$$= \tfrac{1}{2}(2ae^{d/2a} - 2ae^{-d/2a})$$

$$S = ae^{d/2a} - ae^{-d/2a}$$

"And in our case $d = 4$ and $a = 10$," the professor said. We calculated the length of our ribbons:

$$S = (10)(e^{4/20} - e^{-4/20})$$

$$= 4.03$$

"That certainly looks about right," Recordis said. "We know that the ribbons must be a little bit longer than 4, but they can't be very much longer than 4. I'm sure that this is a nice method, but I would feel more comfortable if we used the method to find the length of something that we already know, such as a circle."

We all agreed to that plan. The professor was nervous, though. "After we've come this far, it would be terrible to find that something we've done in calculus is inconsistent."

We set up the equation of a circle:

$$x^2 + y^2 = r^2$$

Using implicit differentiation, we found $dy/dx$:

$$2x + 2y \frac{dy}{dx} = 0$$

$$\frac{dy}{dx} = \frac{-x}{y}$$

$$\left( \frac{dy}{dx} \right)^2 = \frac{x^2}{y^2}$$

$$= \frac{x^2}{r^2 - x^2}$$

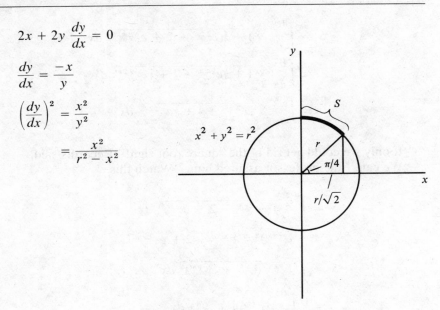

**Figure 14–4.**

We decided to call the length of an arc equal to one eighth of the circumference of the circle $S$ (Figure 14–4).

Next we set up the integral for $S$:

$$S = \int_{x=0}^{x=r/\sqrt{2}} \left[ 1 + \left( \frac{dy}{dx} \right)^2 \right]^{1/2} dx$$

$$= \int_0^{r/\sqrt{2}} \sqrt{1 + [x^2/(r^2 - x^2)]} \ dx$$

$$= \int_0^{r/\sqrt{2}} \sqrt{(r^2 - x^2 + x^2)/(r^2 - x^2)} \ dx$$

$$= \int_0^{r/\sqrt{2}} \sqrt{r^2/(r^2 - x^2)} \ dx$$

$$S = r \int_0^{r/\sqrt{2}} \frac{1}{\sqrt{r^2 - x^2}} \ dx$$

"This is a trigonometric substitution integral," Trigonometeris said.

$$x = r \sin \theta$$

$$dx = r \cos \theta \ d\theta$$

$$\theta = \arcsin \left( \frac{x}{r} \right)$$

We calculated the two limits of integration in terms of $\theta$:

$$\arcsin\left(\frac{1}{\sqrt{2}}\right) = \frac{\pi}{4}$$

$$\arcsin 0 = 0$$

$$S = r \int_0^{\pi/4} \frac{1}{\sqrt{r^2 - r^2 \sin^2\theta}} \, r\cos\theta \, d\theta$$

$$= r \int_0^{\pi/4} \frac{r\cos\theta}{r\cos\theta} \, d\theta$$

$$= r \int_0^{\pi/4} d\theta$$

$$= r\,\theta \Big|_0^{\pi/4}$$

$$= r\left(\frac{\pi}{4} - 0\right)$$

$$S = \frac{\pi r}{4}$$

"Whew! That's right," the professor said. "We know that $C$, the circumference of the whole circle, is given by $C = 8S$, since $S$ is the length of one eighth of the circle. That means $C = 8(\pi r/4) = 2\pi r$. We already know that that is the correct answer."

We went out into the courtyard to cut the ribbons. While we were doing this, Builder came into the courtyard carrying the parabolically shaped clay point-cover, which he prepared to put over the top of Spike Rock.

"You aren't putting it on there like that, are you?" Recordis asked. "That is such an ugly clay color. Don't you think you should paint it first?"

"If you insist," Builder said, although it was clear that he had a lot of other responsibilities as the date of the party approached. "First I have to know how much paint I need."

"It's a good thing he didn't suggest that *we* should figure out how much paint he needs," Recordis remarked, as he returned to ribbon-cutting.

"We should be able to," the king said. "We would have to calculate the surface area of the outside of the paraboloid. Builder is so busy with all the other arrangements that I think we should do this for him."

"We don't know how to calculate surface areas!" Recordis protested. "I hope you aren't about to suggest that we set up a sum that approximates the surface area, and then turn the sum into a definite integral!"

That was exactly what the king had in mind, so we all returned to the Main Conference Room and had Igor draw a picture of a paraboloid (Figure 14–5).

"Now we need to fit some little shapes in there so we can set up a sum," the professor said.

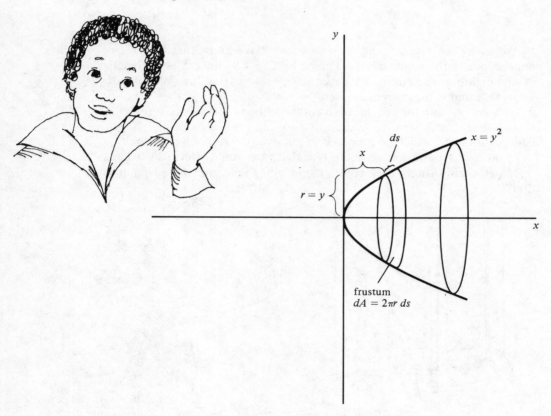

**Figure 14–5.**

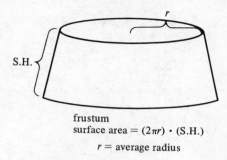

frustum
surface area = $(2\pi r) \cdot$ (S.H.)
$r$ = average radius

**Figure 14–6.**

We decided to fill the paraboloid with a series of sections cut from cones. "I remember what those sections of cones are," Trigonometeris said. "We called them *frustums*. They were very frustrating shapes." (Figure 14–6.)

Recordis looked in his book and found a formula for the surface area of a frustum:

$$\text{(surface area)} = A = 2\pi r \,(\text{slant height})$$

We decided to call the slant height of each frustum $ds$:

$$dA = 2\pi r \, ds$$

"We know what $ds$ is," the professor said. "That's the same $ds$ that we used when we calculated the length of an arc of the curve."

$$ds = \sqrt{1 + (dy/dx)^2} \, dx$$
$$dA = 2\pi r \sqrt{1 + (dy/dx)^2} \, dx$$

"And in the case of the paraboloid the radius of each frustum is simply equal to $y$," the king said.

$$dA = 2\pi y \sqrt{1 + (dy/dx)^2} \, dx$$

The cross section of the paraboloid obeyed the relationship $y = x^{1/2}$, so we could calculate $y$ and $dy/dx$ in terms of $x$:

$$y = x^{1/2}$$

$$\frac{dy}{dx} = \tfrac{1}{2}x^{-1/2}$$

$$\left(\frac{dy}{dx}\right)^2 = \frac{1}{4x}$$

We calculated the surface area for a general paraboloid, having a height of $a$:

$$A = \int_0^a 2\pi\sqrt{x}\sqrt{1 + \frac{1}{4x}}\, dx$$

$$= 2\pi \int_0^a \sqrt{x + \tfrac{1}{4}}\, dx$$

Let $u = x + \tfrac{1}{4};\ du = dx$.

$$A = 2\pi \int_{1/4}^{a+1/4} u^{1/2}\, du$$

$$= 2\pi\, \tfrac{2}{3} u^{3/2} \Big|_{1/4}^{a+1/4}$$

$$= \frac{4\pi}{3}[(a + \tfrac{1}{4})^{3/2} - (\tfrac{1}{4})^{3/2}]$$

"In our case $a = 1$," the king said.

$$A = \frac{4\pi}{3}(1.25^{3/2} - 0.25^{3/2})$$

$$= 5.33$$

"I suppose you want us to calculate the surface area of a sphere, just to make sure that the method works," the professor said to Recordis.

We set up the integral (Figure 14–7).

$$dA = 2\pi y \left[ 1 + \left(\frac{dy}{dx}\right)^2 \right]^{1/2} dx$$

$$\left(\frac{dy}{dx}\right)^2 = \frac{x^2}{r^2 - x^2}$$

$$A = 4\pi \int_0^r \sqrt{r^2 - x^2}\ \sqrt{\frac{r^2 - x^2 + x^2}{r^2 - x^2}}\, dx$$

$$= 4\pi \int_0^r \sqrt{r^2}\, dx$$

$$= 4\pi r \int_0^r dx$$

$$= 4\pi r\, x \Big|_0^r$$

$$= 4\pi r(r - 0)$$

$$A = 4\pi r^2$$

Even Recordis was convinced by now that the method for finding surface areas worked. Just before Builder went out to paint the parabolic point-cover, he asked us whether we could help him with another problem where he thought a continuous sum could be useful.

"I have a large metal half-circle that will be the new concert stage for the party. I want to know the exact point where it will balance so I can

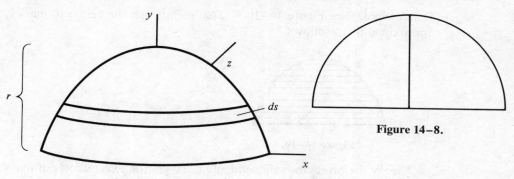

Figure 14–8.

Figure 14–7.

best design the supporting structure to hold it together (Figure 14–8). It is obvious that the balancing point (I call it the *center of mass*) must lie along the center line of the semicircle, because the semicircle is symmetric about the center line. I don't know where to balance it along that line, though."

"How can we help?" the professor asked. "I don't see how that has any connection with what we have been doing."

"We can use your work with discrete sums and continuous sums," Builder said. "It's easy to find the center of mass if you have a certain number of metal bars strung out along a rod." (Figure 14–9.)

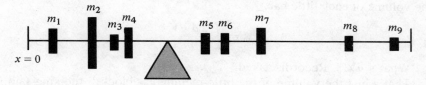

Figure 14–9.

"It doesn't look very easy," the king said. "Suppose each mass was different, or suppose the bars weren't arranged symmetrically."

"You find the center of mass by calculating a *weighted average*."

$$(\text{position of center of mass}) = x_{\text{com}} = \frac{\sum_{i=1}^{n} x_i m_i}{m}$$

"I use $m$ to stand for the total mass of all the bars, and $m_i$ for the mass of each individual bar."

"That tells you where the semicircle balances?" the professor asked, puzzled. (The professor was not very skillful in dealing with practical matters.)

"You bet it does," Builder said.

"How can we figure out the com of the half-circle?" Recordis asked.

"We can imagine that the circle is made up of a bunch of tiny bars,"

Builder said. (See Figure 14–10.) "That means that the center of mass is approximately as follows."

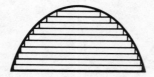

$$x_{\text{com}} = \frac{\displaystyle\sum_{i=1}^{n} x_i m_i}{m}$$

**Figure 14–10.**

"I see!" the professor said suddenly. "To get the exact center of mass you take the limit as the number of bars goes to infinity and the mass of each bar goes to zero!"

$$x_{\text{com}} = \lim_{\Delta m \to 0} \frac{\displaystyle\sum_{i=1}^{n \to \infty} x_i \, \Delta m}{m}$$

"We can write that as an integral!" the king realized.

$$x_{\text{com}} = \frac{\int x \, dm}{m}$$

"What's $dm$?" Recordis asked.

"That must be $dm = \rho \, dV$, where $\rho$ is the density of the plate and $dV$ is the volume of each little bar."

$$x_{\text{com}} = \int \frac{x\rho \, dV}{m}$$

"What's $dV$?" Recordis asked.

"That's just the volume of the little rectangular block," the king said. We decided to call the thickness of each rectangular block $h$ (which is also the thickness of the semicircular plate). The length of each block, we decided, was $2y$, and the width of each block was $dx$ (Figure 14–11). That made the volume of each block equal to $dV = 2yh \, dx$.

$$x_{\text{com}} = \int \frac{x\rho h \, 2y \, dx}{m}$$

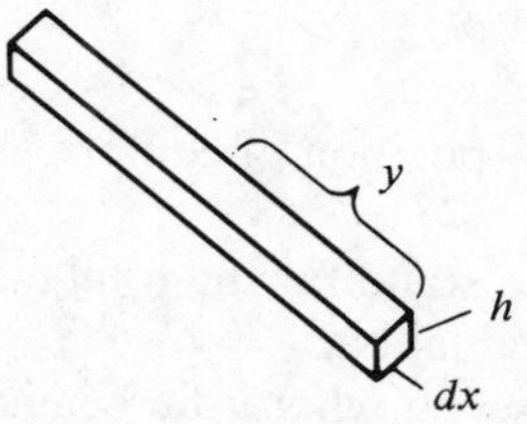

**Figure 14–11.**

"What's $y$?" Recordis asked.
"We can get that from the equation of a circle," the king said.

$$y = \sqrt{r^2 - x^2}$$

$$x_{\text{com}} = 2\rho h \int_0^r \frac{x\sqrt{r^2 - x^2}}{m}\, dx$$

"What's $m$?" Recordis inquired. It was the only other thing he could think of to ask.

"That's easy," Builder said. "It's equal to (volume) × (density)."

$$m = \frac{\rho h \pi r^2}{2}$$

$$x_{\text{com}} = \frac{2\rho h}{\pi \rho h r^2/2} \int_0^r x\sqrt{r^2 - x^2}\, dx$$

"I can take it from here," Recordis interrupted. "I was wondering what was taking you so long."

$$x_{\text{com}} = \frac{4}{\pi r^2} \int_0^r x\sqrt{r^2 - x^2}\, dx$$

Let $u = r^2 - x^2$.

$$du = -2x\, dx$$

$$x_{\text{com}} = \frac{4}{\pi r^2} \left(-\tfrac{1}{2}\right) \int_{r^2}^0 u^{1/2}\, du$$

$$= -\frac{2}{\pi r^2} \tfrac{2}{3} u^{3/2} \Big|_{r^2}^0$$

$$= -\frac{4}{3\pi r^2} \left[0^{3/2} - (r^2)^{3/2}\right]$$

$$= \frac{4r^3}{3\pi r^2}$$

$$= \frac{4r}{3\pi}$$

$$x_{\text{com}} = 0.424r$$

"That looks about right," Builder said. "I would have predicted that the semicircle would have balanced at slightly less than half the radius from the edge."

We went back to work on preparations for the party. That evening the king and I were walking together along the grounds where we had hung the ribbons. "I never would have guessed it," the king said. "We started out trying to solve one simple problem—how fast the train was going at a given instant. Who would have known that we would find a new way to solve optimum-value problems, and a new way for calculating related rates and areas? And now we have the most mysterious part of all—we find that we can add together an infinite number of infinitesimal things."

Exercises

1. Find the length of the arc of the curve $y = x^{3/2}$ from $x = 1$ to $x = 2$.
2. Find the length of the parabola $y = \frac{1}{2}x^2$ from $x = 0$ to $x = a$. (For a helpful integration result, see Chapter 12, exercise 19.) Find a numerical result when $a = 2$.
3. Find the length of the arc of the curve $y = \ln(\sec x)$ from $x = 0$ to $x = \pi/4$. What is the length of the curve from $x = 0$ to $x = \pi/2$?
4. Find the length of the curve $y = ax$ from $x = 0$ to $x = b$.

Set up the integrals for the following arc lengths:
5. $y = x^{1/2}$ from $x = 0$ to $x = a$ $(a > 0)$
6. $y = \ln x$ from $x = 1$ to $x = a$ $(a > 1)$
7. $y = \sin x$ from $x = 0$ to $x = \pi$
8. Find the surface area of the paraboloid $y = x^2$ between $y = 3/4$ and $y = 15/4$.
9. Find the surface area of the cone formed by rotating the line $y = ax$ about the $x$ axis from $x = 0$ to $x = b$.
10. Find the surface area of the solid formed by rotating one arch of the curve $y = \sin x$ about the $x$ axis. (For a helpful integration result, see exercise 2.)
11. Find the surface area of the solid formed by rotating the curve $y = \frac{1}{3}x^3$ about the $x$ axis from $x = 0$ to $x = a$. Find a numerical result when $a = 2$.
12. Find the surface area of the solid formed by rotating the curve $y = e^x$ about the $x$ axis from $x = 0$ to $x = a$.
13. Find the surface area of the top quarter of the sphere of radius $r$:

$$A = 2\pi \int_{r/2}^{r} x\sqrt{1 + \frac{y^2}{x^2}}\, dy$$

14. The ceremonial shield of Carmorra has a thickness of $h$ units and a density of $\rho$, and it is shaped like a parabola $(y = x^2)$. Find the $y$ coordinate of the center of mass $(y_{com})$. The center of mass is located along the line $x = 0$.
15. Find the $y$ coordinate of the center of mass of the cone with cross section $y = ax$ and height $b$.
16. Write the expression for the center of mass of the solid generated by rotating the curve $y = f(x)$ from $x = a$ to $x = b$ about the $x$ axis.
17. The professor has a long rod that is heavier at one end. The cross section of the rod has area $A$, the total length of the rod is $L$, and the density of the rod is given by $\rho = \sqrt{x} + 2$, where $x$ is the distance from the light end of the rod. Find $x_{com}$.
18. The *moment of inertia* of a solid about a particular axis of rotation is given by $I = \int x^2\, dm$, where $x$ is the distance from the axis of rotation. (Note that the moment of inertia for the same solid is different

when different axes of rotation are considered.) For example, a thin rod with cross-sectional area $A$, density $\rho$, and length $L$ has $dm = \rho A\ dx$. Its moment of inertia about an axis through its center perpendicular to the rod (Figure 14–12) is given by:

$$I = \int_{-L/2}^{L/2} x^2 \rho A\ dx = \rho A \int_{-L/2}^{L/2} x^2\ dx = \rho A(\tfrac{1}{3})x^3 \Big|_{-L/2}^{L/2}$$

$$= \frac{\rho A L^3}{12}$$

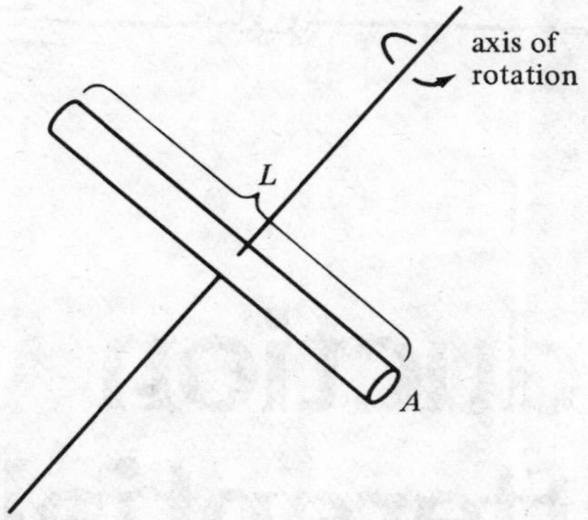

axis of rotation

$L$

$A$

**Figure 14–12.**

Note that $AL = V$, where $V$ is the volume of the rod, so $\rho AL = M$, where $M$ is the total mass of the rod. Therefore $I = ML^2/12$. Find the moment of inertia of the rod about an axis of rotation perpendicular to the rod that passes through one end of the rod.

19. The moment of inertia of a disk (thickness $h$, density $\rho$, radius $R$), for rotations about the axis perpendicular to the disc through its center, is given by:

$$I = \int_{r=0}^{r=R} r^2\ dm$$

where $dm = 2\pi r h \rho\ dr$. Solve the integral to find the moment of inertia. What is $M$, the total mass of the disk? Express the result for $I$ in terms of $M$ and $R$. (One reason that the moment of inertia is important is that the kinetic energy of a body rotating with angular speed $\omega$ is given by K.E. $= \tfrac{1}{2}I\omega^2$.)

20. Find the moment of inertia of a sphere (radius $R$, density $\rho$) with respect to rotations about an axis through its center. (Use cylindrical shells.)

21. Find the moment of inertia of Mongol's conical-shaped top with respect to the central axis of the cone (height $= h$, base radius $= R$).

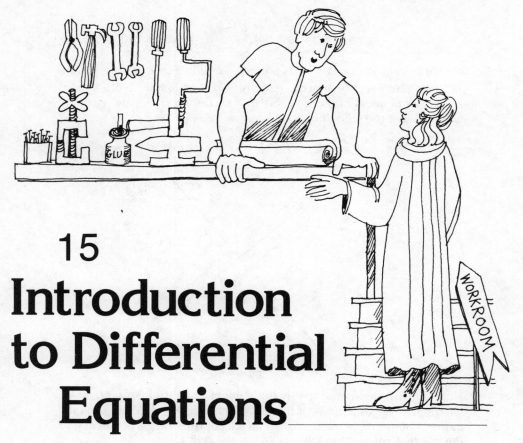

# 15

# Introduction to Differential Equations

Builder was making the final arrangements for the party. "I just thought of something," he said one day. "I have a few old parts in my workroom. I could make them into a nice ride that the children would like. I've figured out the acceleration of the ride, and I'd like to know what its motion looks like."

"How do you know the acceleration?" the professor asked.

"Any object moves according to the equation $F = ma$, where $F$ is the force acting on the object, $m$ is its mass, and $a$ is the acceleration ($a = d^2x/dt^2$). With my ride the force is given by $F = -kx$, and the mass of the ride is $m$. So that makes the final equation of motion as follows."

$$m \frac{d^2x}{dt^2} = -kx$$

"Can you tell me what $x$ is as a function of $t$?"

"We can't do that!" Recordis said. "There's no way to turn that into an integral, and we can solve problems like that only if they're written as integrals."

"If only the equation were $d^2x/dt^2 = -kt$," the professor said. "Then we could integrate the function of $t$ like this."

$$\frac{dx}{dt} = \int(-kt) \, dt, \text{ etc.}$$

192

"If only we had $dx/dt = -kx$," the king pointed out. "Then we could take the $x$ over to the other side."

$$\frac{dx}{x} = -k \ dt, \text{ etc.}$$

"We had better make equations like this one illegal," Recordis said.

"I bet we can find a solution to this kind of problem," the king countered. "The only problem is that we have some function applied to a variable and its derivatives."

$$F\left(t, \ x, \ \frac{dx}{dt}, \frac{d^2x}{dt^2}, \ . \ . \ .\right) = 0$$

"$F$ can be any function of several variables."

"We could call an equation with derivatives in it a *differential equation*," the professor said.

"No way!" Recordis objected. "Take my word for it—as soon as we start applying functions to derivatives, we are really going to be sunk."

"We already know how to solve one type of differential equation," the king said helpfully.

$$\frac{dx}{dt} = f(t)$$

"When we have an equation of that form, we know that the solution is as follows."

$$x = \int f(t) \ dt$$

"But that means that these differential equation things are worse than integrals!" Recordis said. "You're telling me that first we have to solve the differential equation and then we have to solve the integral. I didn't think that anything could be harder than integrals."

"Recordis does have a point," the professor agreed. "We had better put some restrictions on the differential equations before we look for a way to solve them. It looks to me as if it would be very difficult to solve an equation if we are allowed to write an arbitrary function of lots of variables and their derivatives."

"First, we had better say that we are considering only one dependent variable (in this case $x$) and its derivative with respect to one independent variable (in this case $t$)," the king suggested. (A differential equation with one dependent variable and its derivatives with respect to one independent variable is known as an *ordinary differential equation*. If a differential equation involves derivatives with respect to more than one independent variable, it is called a *partial differential equation*.)

"We had better place some more restrictions," the professor said.

"Let's agree not to apply any functions to the derivatives," Recordis said. "We'll be in real trouble if we allow expressions like $(d^2x)^2/dt^2$ or $(d^2x/dt^2)(dx/dt)$. While we're at it, let's make sure that we don't have any functions of $x$, such as $x^2$, or any terms with an $x$ multiplied by its derivative.''

We agreed to these restrictions. They meant that the differential equation could be written in the form:

$$\frac{d^n x}{dt^n} + f_{n-1}(t)\,\frac{d^{n-1}x}{dt^{n-1}} + f_{n-2}(t)\,\frac{d^{n-2}x}{dt^{n-2}} + \cdots + f_0(t)x = f(t)$$

"That looks like a reasonable sort of equation," the professor said. "We have one term with no $x$, one term with one $x$, one term with the first derivative, one term with the second derivative, etc."

Recordis decided to write the equation in a shorter fashion:

$$\left[\frac{d^n}{dt^n} + f_{n-1}(t)\,\frac{d^{n-1}}{dt^{n-1}} + f_{n-2}(t)\,\frac{d^{n-2}}{dt^{n-2}} + \cdots + f_0(t)\right]x = f(t)$$

"Let's refer to this big, long expression with all the derivatives by some letter, say $T$. We can call it a *differential operator*."

$$T = \frac{d^n}{dt^n} + f_{n-1}(t)\,\frac{d^{n-1}}{dt^{n-1}} + f_{n-2}(t)\,\frac{d^{n-2}}{dt^{n-2}} + \cdots + f_0(t)$$

That meant we could write the equation in a much shorter form:

$$Tx = f(t)$$

"I see what an operator is," the professor said. "It's almost like a function. You apply a function to a *number;* the result is another number. You apply an operator to a *function;* the result is another function."

"I know what a differential operator is," the king added. "It means you have a *linear combination* of derivative operators." ("A 'linear combination' is just a fancy way of saying that you set out your list of derivatives, multiply each one by its own coefficient, and then add the whole thing up," Recordis whispered to me.)

We established that any differential operator satisfied the following property:

$$T(ax_1 + bx_2) = aTx_1 + bTx_2$$

where $x_1$ and $x_2$ are two functions and $a$ and $b$ are two constants. We invented the term *linear operator* for an operator satisfying that property.

"Let's call an equation that can be written in this form a *linear differential equation*," the professor said.

---

## LINEAR DIFFERENTIAL EQUATION (GENERAL FORM)

$$\frac{d^n x}{dt^n} + f_{n-1}(t)\,\frac{d^{n-1}x}{dt^{n-1}} + f_{n-2}(t)\,\frac{d^{n-2}x}{dt^{n-2}} + \cdots + f_1(t)\,\frac{dx}{dt} + f_0(t)x = f(t)$$

$Tx = f(t)$     (where $T$ is a differential operator)

"It would help if we got rid of that $f(t)$," the king suggested. "Then we would have an equation where each term contained either $x$ or one of its derivatives."

I suggested that we call this type of equation a *homogeneous linear differential equation*.

---

## HOMOGENEOUS LINEAR DIFFERENTIAL EQUATION (GENERAL FORM)

$$\frac{d^n x}{dt^n} + f_{n-1}(t)\,\frac{d^{n-1}x}{dt^{n-1}} + f_{n-2}(t)\,\frac{d^{n-2}x}{dt^{n-2}} + \cdots + f_1(t)\,\frac{dx}{dt} + f_0(t)x = 0$$

$$Tx = 0$$

---

"That still doesn't look very encouraging," Recordis said. "But it does look as if Builder's equation is linear and homogeneous."

$$\frac{d^2 x}{dt^2} + \frac{k}{m}\,x = 0$$

We spent hours trying to solve this problem. Finally Trigonometeris had an idea. "We could guess," he said.

"We can't guess!" Recordis told him sternly. "This is Math. This is Serious Business. We don't have time for guessing games."

"Where did you get the parts that you are using to build this ride?" Trigonometeris asked Builder.

Builder thought a moment. "I remember," he said. "I'm using the giant spring that was originally part of the gremlin's oscillating chicken-scaring machine."

"Of course!" Trigonometeris said. "Now I know what the answer must be. The chicken-scaring machine moved like a sine wave, so I bet that the answer will be as follows."

$$x = A \sin t$$

"We could try," the professor said. We differentiated Trigonometeris' guess twice to see whether it fit the equation:

$$x = A \sin t$$

$$\frac{dx}{dt} = A \cos t$$

$$\frac{d^2x}{dt^2} = -A \sin t$$

$$= -x$$

"It almost works," the king admitted. He suggested that we try multiplying the $t$ in the middle of the sine function by some number $\omega$:

$$x = A \sin \omega t$$

$$\frac{dx}{dt} = A\omega \cos \omega t$$

$$\frac{d^2x}{dt^2} = -A\omega^2 \sin \omega t$$

$$= -\omega^2 x$$

"Let $\omega = \sqrt{k/m}$," Recordis said. "Then we have $d^2x/dt^2 = (-k/m)x$. Maybe it is possible to solve differential equations after all."

Builder gave us some initial conditions for the ride:

$$t = 0 \text{ when } x = 1; \qquad t = 0 \text{ when } \frac{dx}{dt} = 4$$

We put in the first initial condition to solve for the arbitrary constant $A$:

$$1 = A \sin 0$$

"It won't work!" Recordis moaned. "When $t = 0$ our guess for $x$ will always be zero."

The king had an idea. "Every time we did one integral, we ended up with one arbitrary constant, and we needed one initial condition. And every time we've done two integrals, we've ended up with two arbitrary constants and we've needed two initial conditions. (See the exercises for Chapter 7.) Builder gave us two initial conditions in this case, so I bet that we need to put two arbitrary constants into the solution."

"We put one constant, $A$, in front of the sine function," the professor said. "Where could we put another one?"

"We could put another constant in the middle of the sine," Trigonome-teris suggested.

$$x = A \sin(\omega t + B)$$

We tried this solution and found that we could use our two initial conditions to solve for the two arbitrary constants:

$$1 = A \sin B$$

$$4 = A\omega \cos B$$

$$\tan B = \frac{\omega}{4}$$

$$B = \arctan\left(\frac{\omega}{4}\right)$$

$$A = \frac{1}{\sin B}$$

$$= \frac{\sqrt{16 + \omega^2}}{\omega}$$

Builder was satisfied with this answer, so he left to begin work on the ride.

Encouraged by our success, we decided to look for more ways to solve linear homogeneous differential equations. "We can't guess all the time," the professor said.

$$\frac{d^n x}{dt^n} + f_{n-1}(t)\frac{d^{n-1}x}{dt^{n-1}} + \cdots + f_2(t)\frac{d^2 x}{dt^2} + f_1(t)\frac{dx}{dt} + f_0(t)x = 0$$

"I know another restriction that would help a lot," Recordis offered. "We have all those functions of $t$ in the equation: $f_{n-1}(t)$, etc. It would be much easier if we restricted each derivative to have a constant coefficient. After all, the derivatives in Builder's equation had constant coefficients."

We added that restriction, and then had the general form for a linear homogeneous differential equation with constant coefficients:

$$\frac{d^n x}{dt^n} + c_{n-1}\frac{d^{n-1}x}{dt^{n-1}} + c_{n-2}\frac{d^{n-2}x}{dt^{n-2}} + \cdots + c_2\frac{d^2 x}{dt^2} + c_1\frac{dx}{dt} + c_0 x = 0$$

"I know a function that looks as thought it might fit!" the king said. "The exponential function!"

"Of course!" the professor agreed. "The indestructible exponential function! We know that, if $x = e^t$, then $d^n x/dt^n = x$."

We guessed that the solution to our equation might look like this:

$$x = e^{rt}$$

Then we established that:

$$\frac{dx}{dt} = re^{rt}$$

$$\frac{d^2x}{dt^2} = r^2 e^{rt}$$

$$\frac{d^3x}{dt^3} = r^3 e^{rt}$$

.

.

.

$$\frac{d^n x}{dt^n} = r^n e^{rt}$$

We put that solution back into our equation:

$$r^n e^{rt} + c_{n-1}r^{n-1}e^{rt} + c_{n-2}r^{n-2}e^{rt} + \cdots + c_1 re^{rt} + c_0 e^{rt} = 0$$

"We can factor out the $e^{rt}$," Recordis noted.

$$e^{rt} (r^n + c_{n-1}r^{n-1} + c_{n-2}r^{n-2} + \cdots + c_2 r^2 + c_1 r + c_0) = 0$$

"And no matter what $t$ is, $e^{rt}$ will never be zero," the professor said. "That means that we can divide both sides by $e^{rt}$."

$$r^n + c_{n-1}r^{n-1} + c_{n-2}r^{n-2} + \cdots + c_2 r^2 + c_1 r + c_0 = 0$$

"That's just an algebra equation!" Recordis said with relief. "As much as I like calculus, it is nice to see a regular old algebra equation. Now we can solve for $r$. Of course," he thought a minute more, "it won't be *easy* to solve for $r$, but I'm sure we can do it somehow."

"We could start by restricting our attention to an equation with only two derivatives," the professor said. "That would simplify matters."

We decided that we would call the *order* of a differential equation the highest order derivative that appeared anywhere in the equation. For example, $dx/dt = t^2$ is a first-order equation, $d^2x/dt^2 = -5x$ is a second-order equation, and

$$\frac{d^n x}{dt^n} + f_{n-1}(t) \frac{d^{n-1}x}{dt^{n-1}} + \cdots + f_1(t) \frac{dx}{dt} + f_0(t)x = 0$$

is an equation of order $n$. Igor wrote down the general form for a second-order linear homogeneous differential equation with constant coefficients:

$$\frac{d^2x}{dt^2} + c_1 \frac{dx}{dt} + c_2 x = 0$$

After making the test solution $x = e^{rt}$, we found that $r$ must satisfy this equation:

$$r^2 + c_1 r + c_2 = 0$$

At that moment Recordis began to tremble. Trigonometeris said, "We're forgetting something. Builder's equation was a second-order linear homogeneous differential equation with constant coefficients, but our answer didn't involve an exponent at all. It involved a sine function.''

''Trigonometeris is right,'' the king said. ''This is a mystery.''

We realized that we could solve for the two values of $r$ by using the quadratic formula:

$$r^2 + c_1 r + c_2 = 0$$

$$r = \frac{-c_1 \pm \sqrt{c_1^2 - 4c_2}}{2}$$

''Let's hope that $c_1^2 > 4c_2$,'' Recordis said. ''Then we get two answers:

$$r_1 = \frac{-c_1 - \sqrt{c_1^2 - 4c_2}}{2}$$

$$r_2 = \frac{-c_1 + \sqrt{c_1^2 - 4c_2}}{2}$$

''This way we won't have to use any trigonometry, and we won't have to use any of those . . . those . . . you know, those numbers.'' Recordis still couldn't bring himself to mention the term ''imaginary number.''

''But we did get a trigonometric answer, rather than an exponential answer, to Builder's equation,'' the king said.

''Maybe trigonometric answers and exponential answers are really the same,'' the professor suggested.

''But there is a huge difference,'' the king said. ''An exponential answer gets bigger all the time, whereas a trigonometric answer just oscillates back and forth. We must be able to figure out why we get different kinds of answers.''

We looked at Builder's equation again:

$$\frac{d^2 x}{dt^2} + \omega^2 x = 0$$

(Remember that $\omega^2 = k/m$.)

We tried an exponential solution:

$$x = e^{rt}$$

$$r^2 e^{rt} + \omega^2 e^{rt} = 0$$

$$r^2 + \omega^2 = 0$$

We could solve for $r$ using the quadratic formula, with $c_1 = 0$ and $c_2 = \omega^2$:

$$r = \frac{0 \pm \sqrt{0 - 4\omega^2}}{2}$$

$$= \pm \sqrt{-\omega^2}$$

"No!" Recordis screamed, and fainted. Trigonometeris fanned him absent-mindedly.

"But we can't use that answer, because we don't know what it means when we take the exponential of an imaginary number," the king said. We were able to revive Recordis after we promised him that we would not use any imaginary numbers. (After advancing a bit farther in calculus, we found that it is possible to make a consistent definition for the exponential of an imaginary number. The result is *de Moivre's formula:* $e^{i\theta} = \cos \theta + i \sin \theta$.)

"We can use the imaginary root as a signal that it is time to use a trigonometric answer," the professor suggested. "It looks as though a purely imaginary number will result only when we have an equation of this form."

$$\frac{d^2x}{dt^2} = -c_0 x$$

"If $c_0$ is negative, then we want a function that is proportional to its second derivative with the same sign as its second derivative. That means we use an exponential function. If $c_0$ is positive, that means we need a function that is proportional to its second derivative but has the opposite sign as the second derivative. The one function that satisfies that condition is the function $x = \sin t$."

"Actually there are two functions," Trigonometeris said. "If $x = \cos t$, then $d^2x/dt^2 = -x$."

"All right, we have two functions that work," the professor agreed. "The main thing is that we know we use a trigonometric answer."

"We could use any function similar in form to this one," the king said.

$$x = B_1 \sin \omega t + B_2 \cos \omega t$$

"Remember that the complicated differential operator is a linear operator, so that, if we have two functions, $x_1$ and $x_2$, that satisfy the equation ($Tx_1 = 0$ and $Tx_2 = 0$), we know that any linear combination of the two functions will also be a solution."

$$T(B_1 x_1 + B_2 x_2) = B_1 Tx_1 + B_2 Tx_2 = B_1 0 + B_2 0 = 0$$

"That's just what we want," the professor said. "We have two arbitrary constants, $B_1$ and $B_2$."

"This is getting even more confusing," Recordis said. "When we solved Builder's problem, we said that the answer was $x = A \sin(\omega t + B)$, not $x = B_1 \sin \omega t + B_2 \cos \omega t$."

We puzzled over this mystery for a few minutes before Trigonometeris realized that the two answers are equivalent:

$$x = A \sin(\omega t + B)$$

$$= A \cos B \sin \omega t + A \sin B \cos \omega t$$

Let $A \cos B = B_1,$

$A \sin B = B_2.$

Then

$$x = A \sin(\omega t + B) = B_1 \sin \omega t + B_2 \cos \omega t$$

"We now have the hang of second-order linear homogeneous differential equations with constant coefficients," Recordis said.

$$\frac{d^2x}{dt^2} + c_1 \frac{dx}{dt} + c_0 x = 0$$

"For an equation like that, we set up this equation."

$$r^2 + c_1 r + c_0 = 0$$

"We could call that the *characteristic equation*," the professor suggested.

"Then we solve for $r$," Recordis went on. "If we get two real values of $r$ (call them $r_1$ and $r_2$), we set up this solution."

$$x = B_1 e^{r_1 t} + B_2 e^{r_2 t}$$

"Then we use the two initial conditions to solve for the two arbitrary constants, $B_1$ and $B_2$. If we get two imaginary values for $r$ (call them $i\omega$ and $-i\omega$), we just ignore the fact that the numbers are imaginary and set up the solution for $x$."

$$x = B_1 \sin \omega t + B_2 \cos \omega t$$

"We can rewrite this using one sine function if we want to. That should take care of everything except . . ." Recordis almost fainted again. "What if we have a complex solution, such as $r = r_0 + i\omega$?"

At that moment Builder came into the room. "I left out an important term in the equation I gave you," he said. "When I was building the ride, I found that there is a certain amount of friction that acts to slow the ride down. The faster the ride goes, the greater is the force of the friction. The equation of motion for the ride should be as follows."

$$\frac{d^2x}{dt^2} = -b \frac{dx}{dt} - \omega^2 x$$

"Here $b$ is a constant that measures the friction," Builder finished. We rewrote the equation:

$$\frac{d^2x}{dt^2} + b \frac{dx}{dt} + \omega^2 x = 0$$

Then we set up the characteristic equation:

$$r^2 + br + \omega^2 = 0$$

$$r = \frac{-b \pm \sqrt{b^2 - 4\omega^2}}{2}$$

Builder gave us some numerical values: $b = 3$, $\omega^2 = 5$.

$$r = \frac{-3 \pm \sqrt{9 - 20}}{2}$$

$$= \frac{-3}{2} \pm \tfrac{1}{2}\sqrt{-11}$$

$$= \frac{-3}{2} \pm i\,\tfrac{1}{2}\sqrt{11}$$

$$r_1 = -1.5 + 1.66i$$

$$r_2 = -1.5 - 1.66i$$

"We're in real trouble now!" Recordis said. "If we had a purely real number, we could use an exponential solution. If we had a purely imaginary number, we could use a trigonometric solution. But we have a complex number: partly real and partly imaginary."

"We could try both kinds of solutions," the king guessed. "We could put the real part in an exponential solution, and we could put the imaginary part in a trigonometric solution."

The king guessed the following solution:

$$x = e^{-1.5t}\,[B_1 \sin(1.66t) + B_2 \cos(1.66t)]$$

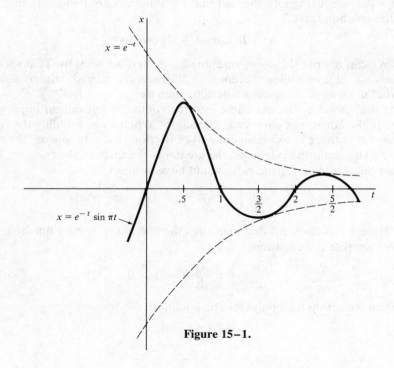

Figure 15–1.

"What does that function look like?" Recordis asked. "I never saw a solution like that before."

Igor drew a graph of the curve $x = e^{-t} \sin t\pi$ (Figure 15–1).

"That's right," the professor said. "The solution keeps going back and forth, as a ride on a spring should, but the amplitude of each swing is becoming less. The friction acts to damp out the sine wave."

"I could have told you that is what the motion of the ride would look like before we developed all this math," Builder said. "All you need is some physical intuition."

"We could call that kind of curve a *damped sine wave,*" Recordis suggested. "But we better make sure that it is the right answer."

We differentiated our trial solution and put it back into Builder's equation. The first derivative required the product rule:

$$x = e^{-1.5t}[A \sin(1.66t + B)]$$

$$\frac{dx}{dt} = e^{-1.5t} A\ 1.66 \cos(1.66t + B) - 1.5e^{-1.5t} A \sin(1.66t + B)$$

We expressed this answer in slightly simpler form:

$$\frac{dx}{dt} = (Ae^{-1.5t})(1.66 \cos \theta - 1.5 \sin \theta)$$

(To save us some writing, we made the definition $\theta = 1.66t + B$.)

$$\begin{aligned}
\frac{d^2x}{dt^2} &= (Ae^{-1.5t})(-1.66^2 \sin \theta - 1.5 \times 1.66 \cos \theta) \\
&\quad - (1.5Ae^{-1.5t})(1.66 \cos \theta - 1.5 \sin \theta) \\
&= (1.5^2 - 1.66^2)Ae^{-1.5t} \sin \theta - 2(1.5 \times 1.66)Ae^{-1.5t} \cos \theta \\
&= -0.5Ae^{-1.5t} \sin \theta - 4.98Ae^{-1.5t} \cos \theta
\end{aligned}$$

We recognized our expression for $x$:

$$x = e^{-1.5t} A \sin \theta$$

so we could rewrite the expression for the second derivative:

$$\frac{d^2x}{dt^2} = -0.5x - 4.98Ae^{-1.5t} \cos \theta$$

Now that we knew $d^2x/dt^2$, $dx/dt$, and $x$, we could put all the parts together in Builder's equation and see if they fit.

We found an expression for 3 $dx/dt$:

$$\frac{dx}{dt} = (Ae^{-1.5t})(1.66 \cos \theta - 1.5 \sin \theta)$$

$$= 1.66Ae^{-1.5t} \cos \theta - 1.5x$$

$$3\ \frac{dx}{dt} = 4.98Ae^{-1.5t} \cos \theta - 4.5x$$

Then we put all the parts together:

$$\frac{d^2x}{dt^2} + 3\,\frac{dx}{dt} + 5x = -0.5x - 4.98Ae^{-1.5t}\cos\theta + 4.98Ae^{-1.5t}\cos\theta$$
$$- 4.5x + 5x$$
$$= -0.5x - 4.5x + 5x$$
$$= 0$$

"It does work!" Trigonometeris said. "We do need both an exponential solution and a trigonometric solution for that type of problem!" We found that we could solve for the arbitrary constants $A$ and $B$ by using the initial conditions that Builder had given us. We wrote down a general procedure for solving second-order linear homogeneous differential equations with constant coefficients:

## SECOND-ORDER LINEAR HOMOGENEOUS CONSTANT-COEFFICIENT DIFFERENTIAL EQUATION SOLUTION METHOD

1. Make sure that the equation is ordinary (there is only one dependent variable and its derivatives with respect to only one independent variable).
2. Make sure that the equation is second order, that is, no derivatives other than first or second derivatives appear in the equation.
3. Make sure that the equation is linear, that is, it can be written in the form:

$$\frac{d^2x}{dt^2} + f_1(t)\,\frac{dx}{dt} + f_0(t)x = f(t)$$

4. Make sure that the equation is homogeneous, that is, each term contains an $x$ or one of its derivatives, so it can be written in the form:

$$\frac{d^2x}{dt^2} + f_1(t)\,\frac{dx}{dt} + f_0(t)x = 0$$

(Also, remember that in this chapter we have used $x$ as the dependent variable and $t$ as the independent variable!)

5. Make sure that $x$ and all its derivatives have constant coefficients:

$$\frac{d^2x}{dt^2} + c_1\,\frac{dx}{dt} + c_2x = 0$$

6. Next, set up the characteristic equation:

$$r^2 + c_1r + c_2 = 0$$

"Solving differential equations is pretty complicated, but it may not be as bad as I thought it would be," Recordis said.

"I see that we have the solution," Builder added, "but that solution does not leave us with a very good ride. The children won't want a ride that just swings back and forth for a little bit before it damps out and stops. I'll put a driving motor on the ride so that it won't stop so quickly."

Before we could protest, Builder had designed a driving motor that would push the spring ride with a force equal to $F_{driving} = D \sin \Omega t$, where $\Omega$ is the driving angular frequency.

"No!" Recordis cried. "Now we won't have a homogeneous equation any more! That driving force term does not have any $x$'s in it."

Builder gave us the equation for the driven ride:

$$\frac{d^2x}{dt^2} + b\,\frac{dx}{dt} + \omega^2 x = D \sin \Omega t$$

---

7. Solve for $r$, using the quadratic formula:

$$r = \frac{-c_1 \pm \sqrt{c_1{}^2 - 4c_2}}{2}$$

8. If $r$ has two real values ($r_1$ and $r_2$), set up an exponential solution:

$$x = B_1 e^{r_1 t} + B_2 e^{r_2 t}$$

9. If $r$ has one real value ($r_0$), set up a solution as follows:

$$x = B_1 e^{r_0 t} + B_2 t e^{r_0 t}$$

(See exercise 14.)

10. If $r$ has two purely imaginary values ($i\omega$ and $-i\omega$), set up the trigonometric solution:

$$x = B_1 \sin \omega t + B_2 \cos \omega t$$

(or $x = A \sin(\omega t + B)$)

11. If $r$ has two complex values ($r_0 + i\omega$ and $r_0 - i\omega$), set up a solution as follows:

$$x = e^{r_0 t}(B_1 \sin \omega t + B_2 \cos \omega t)$$

12. In all of these cases the solution contains two arbitrary constants. If you have two initial conditions, you can solve for the arbitrary constants and complete the solution.

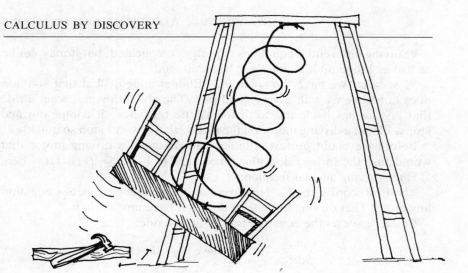

"That's a second-order linear differential equation with constant coefficients, but it's not homogeneous," the professor objected. "How are we going to solve that?"

"The solution still must involve a sine function," Trigonometeris said. "The ride will still go back and forth, even with the driving force."

Trigonometeris guessed that the frequency of the ride should be close to the frequency of the driving force, so we guessed the following solution:

$$x = C \sin(\Omega t + B)$$

$$\frac{dx}{dt} = \Omega C \cos(\Omega t + B)$$

$$\frac{d^2x}{dt^2} = -\Omega^2 C \sin(\Omega t + B)$$

We put these expressions back into Builder's equation:

$$\frac{d^2x}{dt^2} + b\frac{dx}{dt} + \omega^2 x = -\Omega^2 C \sin(\Omega t + B) + b\Omega C \cos(\Omega t + B)$$
$$+ \omega^2 C \sin(\Omega t + B)$$
$$= (\omega^2 - \Omega^2)C \sin(\Omega t + B) + b\Omega C \cos(\Omega t + B)$$

Using the formula for the sine and cosine of the sum of two angles gave us:

$$\frac{d^2x}{dt^2} + b\frac{dx}{dt} + \omega^2 x = (\omega^2 - \Omega^2)\,C(\sin \Omega t \cos B + \cos \Omega t \sin B)$$
$$+ b\Omega C(\cos \Omega t \cos B - \sin \Omega t \sin B)$$

We simplified this expression a bit:

$$\frac{d^2x}{dt^2} + b\frac{dx}{dt} + \omega^2 x = [C \cos B(\omega^2 - \Omega^2) - b\Omega C \sin B] \sin \Omega t$$
$$+ [C \sin B(\omega^2 - \Omega^2) + b\Omega C \cos B] \cos \Omega t$$

INTRODUCTION TO DIFFERENTIAL EQUATIONS **207**

Now we had the left-hand side of Builder's equation for the damped-driven ride. The next step was to set our equation for the left-hand side equal to the right-hand side of the equation:

(right-hand side) $= D \sin \Omega t$

$$(\text{left-hand side}) = [C \cos B(\omega^2 - \Omega^2) - b\Omega C \sin B] \sin \Omega t$$
$$+ [C \sin B(\omega^2 - \Omega^2) + b\Omega C \cos B] \cos \Omega t$$
$$[C \cos B(\omega^2 - \Omega^2) - b\Omega C \sin B] \sin \Omega t$$
$$+ [C \sin B(\omega^2 - \Omega^2) + b\Omega C \cos B] \cos \Omega t$$
$$= D \sin \Omega t$$

If our alleged solution really worked, we wanted it to be true for all values of $t$. That meant that we could set the coefficient of $\sin \Omega t$ on the left equal to the coefficient of $\sin \Omega t$ on the right:

coefficients of $\sin \Omega t$:

$$D = C \cos B(\omega^2 - \Omega^2) - b\Omega C \sin B$$

coefficients of $\cos \Omega t$:

$$0 = C \sin B(\omega^2 - \Omega^2) + b\Omega C \cos B$$

We could find the conditions that $B$ and $C$ must satisfy for these two equations to be correct:

$$0 = (\omega^2 - \Omega^2) \sin B + b\Omega \cos B$$

$$(\Omega^2 - \omega^2) \sin B = b\Omega \cos B$$

$$\tan B = \frac{b\Omega}{\Omega^2 - \omega^2}$$

$$B = \arctan\left(\frac{b\Omega}{\Omega^2 - \omega^2}\right)$$

$C$ must satisfy the equations:

$$C = \frac{D}{\cos B(\omega^2 - \Omega^2) - b\Omega \sin B}$$

$$\sin B = \frac{b\Omega}{\sqrt{b^2\Omega^2 + (\Omega^2 - \omega^2)^2}}$$

$$\cos B = \frac{\Omega^2 - \omega^2}{\sqrt{b^2\Omega^2 + (\Omega^2 - \omega^2)^2}}$$

$$C = \frac{D}{\dfrac{\Omega^2 - \omega^2}{\sqrt{b^2\Omega^2 + (\Omega^2 - \omega^2)^2}}(\omega^2 - \Omega^2) - b\Omega \dfrac{b\Omega}{\sqrt{b^2\Omega^2 + (\Omega^2 - \omega^2)^2}}}$$

$$= -\frac{D}{\dfrac{(\Omega^2 - \omega^2)^2 + b^2\Omega^2}{\sqrt{b^2\Omega^2 + (\Omega^2 - \omega^2)^2}}}$$

We were all exhausted at this point, but Recordis suddenly exclaimed, ''We can simplify that!''

$$C = -\frac{D}{\sqrt{b^2\Omega^2 + (\Omega^2 - \omega^2)^2}}$$

We put together the final answer:

$$x = C \sin(\Omega t + B)$$

$$B = \arctan\left(\frac{b\Omega}{\Omega^2 - \omega^2}\right)$$

$$C = -\frac{D}{\sqrt{b^2\Omega^2 + (\Omega^2 - \omega^2)^2}}$$

''We've come up with an answer,'' Recordis said with relief. ''Now all we have to do is put in the numbers. And this time we were able to solve for all the constants, rather than . . .'' He suddenly stopped.

''Something is wrong!'' the professor said. ''We don't have any arbitrary constants left! We'll never be able to match the initial conditions now. We know that, whenever we have a second-order differential equation, the solution must contain two arbitrary constants.''

We were stumped. ''That solution doesn't have any place left to put any constants,'' Recordis complained.

We thought for a long time. ''We'll have to add something to that function,'' I suggested.

$$x = x_1 + x_2$$

$$x_1 = C \sin(\Omega t + B)$$

$$x_2 = ?$$

''Whatever $x_2$ is, it must be some function that has two arbitrary constants in it.''

''But that will wreck our solution to the equation,'' Recordis protested. ''You can't just add any old thing to the solution and still expect to have a solution.''

''Maybe we can find some special $x_2$ that will be all right,'' the king said hopefully. ''Maybe there is one $x_2$ such that $x = x_1 + x_2$ will still be a solution.''

We put this alleged solution back into the equation (written in operator form):

$$\left(\frac{d^2}{dt^2} + b\,\frac{d}{dt} + \omega^2\right)(x_1 + x_2) = D \sin \Omega t$$

$$T(x_1 + x_2) = D \sin \Omega t$$

''We know that

$$T = \frac{d^2}{dt^2} + b\,\frac{d}{dt} + \omega^2$$

is a linear operator," the professor said. "We can rewrite our equation like this."

$$Tx_1 + Tx_2 = D \sin \Omega t$$

"We know that $Tx_1 = D \sin \Omega t$, since that is the right answer for the equation that we just found."

$$D \sin \Omega t + Tx_2 = D \sin \Omega t$$

$$Tx_2 = 0$$

"That puts a restriction on $x_2$," the professor said. "We can safely add $x_2$ to our original solution without wrecking it if the following equality holds."

$$\left(\frac{d^2}{dt^2} + b\,\frac{d}{dt} + \omega^2\right)x_2 = 0$$

"Does anybody see any clues in that equation that will allow us to determine what $x_2$ is?"

We stared hard at the last equation. "We already solved that equation!" the king exclaimed suddenly. "That's just the homogeneous equation we had before Builder added the driving machine!"

"That means that $x_2$ is just the homogeneous solution," Recordis said in surprise.

$$x_2 = e^{-1.5t}\,A_1 \sin(1.66t + A_2)$$

"It even has two arbitrary constants, just as it is supposed to."

We wrote out the whole solution:

$$\frac{d^2x}{dt^2} + b\,\frac{dx}{dt} + \omega^2 x = D \sin \Omega t$$

$$x = C \sin(\Omega t + B) + e^{r_0 t} A_1 \sin(r_1 t + A_2)$$

$$C = -\frac{D}{\sqrt{b^2\Omega^2 + (\Omega^2 - \omega^2)^2}}$$

$$B = \arctan\left(\frac{b\Omega}{\Omega^2 - \omega^2}\right)$$

$$r_0 = \frac{-b}{2}$$

$$r_1 = \frac{\sqrt{4\omega^2 - b^2}}{2}$$

$A_1$ and $A_2$ are arbitrary constants.

(Note that this solution holds when $4\omega^2 > b^2$.)

"What does that solution look like?" Recordis asked. "It looks as though it will be very complicated."

"It will be very complicated at first," Builder said thoughtfully. "But after a fair amount of time has passed, it looks as if we will be able to completely ignore the second term, since it will all damp out."

"That's right," the professor agreed. "As $t$ becomes large, $e^{-1.5t}$ will go to zero. Then the only part of the solution we will have to worry about will be the first part.

"We could say that the $x_2$ part, which comes from the homogeneous equation solution, is a *transient* part of the solution. If we wait long enough, it will go away."

"The $x_1$ part will be a permanent part of the solution," Recordis said. "As $t$ becomes large, the ride will keep oscillating. And that part doesn't even depend on the initial conditions."

"That makes sense," Builder concurred. "At first, the solution should depend a lot on the initial conditions. As time goes by, though, the effects of the initial conditions will be less important, until ultimately it will be only the driving force that makes a difference."

After Builder had worked on his drawings for a few moments, he noticed something else. "It is a good thing that the friction is there," he said. "Suppose there was no friction ($b = 0$). Then the permanent solution would be as follows."

$$x = \frac{D}{\Omega^2 - \omega^2} \sin \Omega t$$

"I am trying to figure out the driving frequency that would give the ride the maximum amplitude. It looks as though the maximum amplitude would occur if the driving frequency $\Omega$ equals $\omega$. But if that happens, the amplitude of the ride will be infinity, which means that it will shoot all over space. The children would probably think that was a fun ride, but I don't think it would be very safe."

"I remember what $\omega$ is," the king said. "That was the angular frequency of the ride before you added the driving motor. We could call that the *natural frequency* of the ride. It looks as though the most amplitude occurs if the frequency of the driving motor is close to the natural frequency of the ride." We decided to call this occurrence *resonance*.

"Fortunately, the friction is there," Builder noted. "That means that the ride will not go off to infinity, even if we use the resonant driving frequency."

"I've had enough of differential equations," Recordis said. "It's time to go back to work on the party."

We wrote down a general procedure for solving linear differential equations:

## LINEAR DIFFERENTIAL EQUATION SOLUTION METHOD

1. Check to see whether the equation is homogeneous. If it is not homogeneous, pretend for the moment that it is.
2. Find the solution of the resulting homogeneous equation. If the equation is second order and has constant coefficients, you can use the method outlined earlier in this chapter. In any case, your solution should have as many arbitrary constants as the order of the equation. For example, if you have a second-order equation, the homogeneous solution should have two arbitrary constants.
3. Next, find any function that acts as a solution of the actual equation. (You have to stop pretending that the equation is homogeneous now.) We call this the *particular* solution. If you don't recognize the equation, you will have to keep guessing until you find the right particular solution.
4. The final solution is the sum of the homogeneous solution and the particular solution.
5. Use your initial conditions to solve for the arbitrary constants.

"We'd better not advertise that we can solve differential equations yet," the professor said, "or someone may come to us with an equation that is too complicated for us to solve."

We were all glad that we were done for the day. As we were leaving to finish the preparations for the party, the king said to me, "Everything seems to be going perfectly. There is one thing that worries me, though."

"What's that?" I asked.

"We haven't heard from the gremlin lately, and it is unlike him to be quiet this long."

## Exercises

Which of these differential equations are linear? Which are homogeneous?

**1.** $\dfrac{d^2y}{dx^2}\, y^2 + 2\, \dfrac{dy}{dx} = 0$

**2.** $\dfrac{dy}{dx} = f(x)$

**3.** $3\left(\dfrac{dy}{dx}\right)^2 + 4y = 5x$

**4.** $\dfrac{dy}{dx} = f(y)$

**5.** $x\, \dfrac{dy}{dx} - 3x^2 + 2y = 4\, \dfrac{dy}{dx}$

**6.** $\dfrac{d^2x}{dt^2}\, \dfrac{dx}{dt} = 4t$

**7.** $\dfrac{d^2y/dx^2}{[1 + (dy/dx)^2]^{1.5}} = K$

**8.** For

$$T = \dfrac{d^n}{dt^n} + f_{n-1}(t)\, \dfrac{d^{n-1}}{dt^{n-1}} + \cdots + f_2(t)\, \dfrac{d^2}{dt^2} + f_1(t)\, \dfrac{d}{dt} + f_0(t)$$

show that $T(ax_1 + bx_2) = aTx_1 + bTx_2$ ($a$ and $b$ are constants; $x_1$ and $x_2$ are functions of $t$).

Solve these differential equations:

**9.** $\dfrac{d^2x}{dt^2} + \dfrac{dx}{dt} - 6x = 0$

**10.** $\dfrac{d^2x}{dt^2} + \dfrac{dx}{dt} + x = 0$

**11.** $\dfrac{d^2x}{dt^2} + 9x = 0$

**12.** $\dfrac{d^2x}{dt^2} + 2\, \dfrac{dx}{dt} + x = 0$

**13.** $\dfrac{d^2x}{dt^2} - 4\, \dfrac{dx}{dt} = 0$

**14.** Step 9 of the method for solving a homogeneous equation says that, if the characteristic equation has one real root ($r$), the solution will be $x = Ae^{rt} + Bte^{rt}$. Differentiate this solution to show that it is the correct one.

15. Find the motion of Builder's ride if $b = 3$, $\omega^2 = 5$, $\Omega = 2$, and $D = -4$.

16. Solve for the arbitrary constants in the solution to exercise 15 by using these initial conditions: when $t = 0$, $x = -0.6484$ and $dx/dt = 1.884$.

17. Consider an undriven ride with greater friction. Let $\omega = 2$, $b = 4$. What does the motion of the ride look like?

18. Consider another undriven ride with an even greater force of friction: $b = 5$ and $\omega = 2$. What does the motion of the ride look like?

# 16
# Comprehensive
# Test
# of Calculus
# Problems

Two days before the party, when everyone was in the Main Conference Room, there was a knock at the door, and we opened it to find, to our surprise, the gremlin. He was dressed in the same evil-looking cape that he always wore, but this time he had not come flying in the window and he had toned down the cackle in his laugh. He was carrying a large scroll, which he tossed on the table.

"Take that!" he commanded.

"Thank you," Recordis said. "We appreciate this very much. What is it?"

"Fool!" He slyly slithered over to the trembling professor. "So you are intending to write a book, are you? Before you can do that, you must solve these. These are all problems—calculus problems. If you can do them all, then you will have something to write about. But I think you will not be able to." He turned to the king. "There is no gimmick this time. I will not engulf the kingdom in fire or anything else if (I mean *when*) you fail. I shall simply proclaim to the people at the party that you, who have spent so much time on this calculus, cannot even solve calculus problems. The people of Carmorra themselves will turn against you, and I shall be proclaimed king!"

"We've defeated you every other time so far," Recordis said. "Your win-loss record is 0-17."

"Ah! But I realize where I have been going wrong," the gremlin said. "I have only fought you one at a time on theoretical grounds. Theory, unfortunately, is your strong point. Now you must apply everything you have done. That is where I think you will fail."

He reached inside his cloak and pulled out a giant conical-shaped hourglass. "The sand in this glass has been set to run for 48 hours. Once it is empty, I shall return and witness your defeat. I hope you appreciate the trouble I had rounding up these problems. It has been fun knowing you!" With a sudden burst of his horrible laugh he swept out the window.

"We surely got rid of him!" Recordis said.

"What do you mean, 'we got rid of him?' " the professor exclaimed. "We still have to do all these problems before he comes back!"

"Since you're writing the book, we'll let you do the problems for us," Recordis said, packing his notebooks and heading for the door.

"Stop it!" the king commanded. "We have no time for any of this arguing. We all helped get us into this, and we're all going to help get us out."

There was a long silence while we contemplated the scroll on the table. Mongol began to cry again.

"We may as well get started," the professor said. Recordis began reading through the 45 problems that were our final test:

1. Find the volume of the top half of the hourglass that is even now measuring the time until your doom. (The hourglass is a cone: the radius of the base is $r = 15$, and the height is $h = 30$.)

2. The sand falls through the opening in the hourglass at the rate of $u$ cubic units per unit time. What is the rate of change in the height of the sand in the cone when $x = 5$? ($u = 0.04$.) (See Figure 16–1.)

3. You have 48 hours to solve 45 problems. I predict that the rate at which you solve problems will be given by:

$$\frac{dn}{dt} = 181e^{-2t} - 1$$

where $n$ is the number of problems you have solved and $t$ is the time measured in hours ($t = 0$ at the time you start working on this test). If you follow my prediction, how many problems will you have finished by the time $t = 48$?

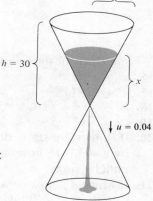

Find the derivative of the following functions:

4. $f(x) = x^3 - 3x^2 + 4x + 10$
5. $f(x) = x^{96} + 10x^{43} - 3x^{20} + 11x$
6. $f(x) = x^2 + x + 1 + 1/x + 1/x^2 + 1/x^3$
7. $f(x) = \sqrt{4 - x^3}$

**Figure 16–1.**

8. $f(x) = (10 + x^4)/(3 - x^2)$
9. $f(x) = [3 + (x + 4)^3]^2$
10. $f(x) = x^3 \sqrt{3 + x^2}$
11. $f(x) = \sqrt{(x + 5)(x - 3)/(x + 2)(x - 1)}$
12. $f(x) = \sin^2 x + \cos^2 x$
13. $f(x) = x^2 e^x \sin(x) \ln x$
14. $f(x) = e^{\ln x}$
15. When you are planning your escape from the kingdom, you will want to reach Nowhere Island as quickly as possible. Nowhere Island is $a = 10$ km directly offshore from River Mouth. When you arrive at Shore Rock, you will be $b = 12$ km away from River Mouth. Assume that you are carrying your boat with you. You can row $v = 3$ km/hr, and you can walk $u = 5$ km/hr. What course should you follow to arrive at Nowhere Island as quickly as possible? (See Figure 16–2.)

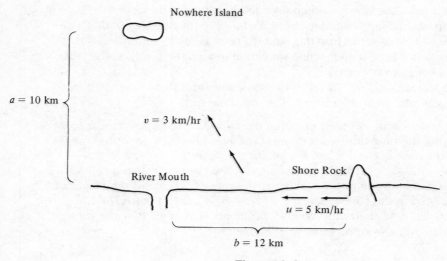

**Figure 16–2.**

Find $dy/dx$ for the following relations ($r$, $a$, and $b$ are constants):
16. $x^2 + y^2 = r^2$
17. $x^2/a^2 + y^2/b^2 = 1$
18. $(x + y)^{1/2} = (x - y)^{-1/2}$
19. $(x - 3)y^2 = x^2 + 4$
20. If Rutherford starts running on a block of ice where his speed is given by $v = 3 - t^2$, how far will he have traveled from the time when $t = 0$ until he stops?
21. If Mongol jumps on a muddy hillside where his speed changes at the rate $dv/dt = -2v$, with $v = 5$ when $t = 0$, how long will it take until his speed is equal to 0.01, and how far will he have traveled by the time this happens?

Find the values of the following integrals:

**22.** $\int (4x^3 - 2x + 5)\, dx$

**23.** $\int x\, \sqrt{x^2 - 1}\, dx$

**24.** $\displaystyle\int \frac{1}{1 + x^2}\, dx$

**25.** $\int 4x^2 \sin^2 x\, dx$

Find the equation for $y$, given the following conditions:

**26.** $dy/dx = 5x + 4;$      $y = 3$ when $x = 2$

**27.** $d^2y/dt^2 = 5;$      $dy/dt = 3$ and $y = 2$ when $t = 0$

**28.** $d^2y/dt^2 = 2t;$      $dy/dt = 0$ and $y = 0$ when $t = 0$

**29.** Mongol went for a ride on a spring where he was acted upon by the following forces: $F_{spring} = -500x;$ $F_{friction} = -10\, dx/dt$. The mass of Mongol is 1000. At time $t = 0$, his speed was zero ($dx/dt = 0$) and his position was 20 ($x = 20$). Set up the differential equation that represents his motion, and solve it to find $x$ as a function of $t$.

**30.** Suppose that the *work* required to move Irving Electron from point $a$ to point $b$ against a force $F$ is given by:

$$W = \int_a^b F(x)\, dx$$

What is the work if $a = 100$, $b = 10$, and $F(x) = -kx^{-2}$?

Find the values of these definite integrals:

**31.** $\displaystyle\int_4^5 \frac{x - 1}{(x - 2)(x - 3)}\, dx$

**32.** $\displaystyle\int_0^{10} 4x^2\, dx$

**33.** $\displaystyle\int_0^{\pi/6} \tan 2x\, dx$

**34.** $\displaystyle\int_0^{10} (e^x - x^3)\, dx$

**35.** $\displaystyle\int_1^2 x^4 \sin(x^5)\, dx$

**36.** $\displaystyle\int_0^{\pi/2} \sin x \cos x\, dx$

**37.** $\displaystyle\int_2^3 x^2 \sin x\, dx$

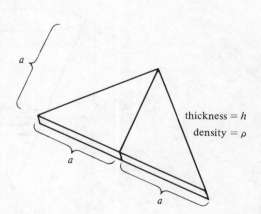

**Figure 16–3.**

**38.** Find the area between the curves $y_1 = \sin x$ and $y_2 = x^2 - \pi x$.

**39.** When I take over the kingdom, I will want a new concert hall designed with a triangular stage. (See Figure 16–3.) Find the point where it will balance (the center of mass).

**40.** If a football is formed by rotating the sine curve $y = \sin x$ from zero to $\pi$ about the $x$ axis, find the volume of the football.

**41.** Sketch a graph of the curve $y = e^{-x^2}$. Make a list of places with horizontal tangents and points of inflection.

**42.** Solve the differential equation $d^2y/dx^2 + 2\, dy/dx + 2y = 10$.

**43.** The two towers of the Swirling Pass Suspension Bridge are 100 units apart. The roadway is 40 units below the top of the towers. The main cable of the bridge hangs freely between the two towers, with its lowest point just touching the roadway. What is the length of the main cable? (Equation of main cable: $y = 18(e^{x/36} + e^{-x/36}) - 36$.)

**44.** Find the area between the curves $y_1 = x^2 - 5$ and $y_2 = x - 5$.

**45.** Find the derivative of the function $y = |x|$ ($y$ is the absolute value of $x$, defined by $|x| = x$ for $x \geq 0$ and $|x| = -x$ for $x < 0$) at the point where $x = 0$.

"Some of these are easy, and some are hard," Recordis said. "Which should we do first?"

"We'll split up," the king told him. We broke into teams and set to work on the problems. The next two days passed in a blur. The only things we saw were problems. I made no effort to keep track of the steps we took chronologically, but I did make a record of the solutions.

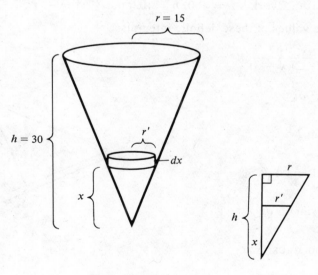

**Figure 16–4.**

**1.** We divided the hourglass into little cylindrical disks (Figure 16–4). The volume of each cylinder is given by $dV = \pi r'^2\, dx$, and $r'$ can be found using similar triangles: $r'/x = r/h$. So $r' = rx/h$. Then we integrated from $x = 0$ to $x = h$:

$$\int dV = V = \int_0^h \frac{\pi r^2 x^2 \, dx}{h^2} = \frac{\pi r^2}{h^2} \tfrac{1}{3} x^3 \Big|_0^h = \frac{\pi r^2 h}{3}$$

"That formula should be good for any cone," the professor noted. Recordis inserted the numbers: $V = 7069$.

"That scoundrel!" Recordis cried. "He made the hourglass just small enough!"

**2.** The first thing the professor suggested was that we write down everything that we knew:

$$\frac{dv}{dt} = 0.04$$

($v$ is the volume of sand left in the top of the hourglass at time $t$).

$$\frac{r'}{r} = \frac{x}{h}$$

(from a similar triangle relationship).

Now we were stuck, until we realized that we needed the result from problem 1 so that we knew the volume of a cone:

$$v = \frac{\pi r'^2 x}{3}$$

The professor noticed that we could substitute $r' = rx/h$:

$$v = \frac{\pi r^2 x^3}{3h^2}$$

Now we could determine $dv/dx$:

$$\frac{dv}{dx} = \frac{\pi r^2 x^2}{h^2}$$

The problem was to find $dx/dt$. We knew $dv/dx$ and $dv/dt$. The king realized that we could use the chain rule:

$$\frac{dv/dt}{dv/dx} = \frac{dv}{dt}\frac{dx}{dv} = \frac{dx}{dt}$$

$$\frac{dx}{dt} = \frac{0.04h^2}{\pi r^2 x^2}$$

This expression told us the rate at which the height of the sand was changing for any value of the height $x$, but the gremlin asked us only for the answer when $x = 5$. We put these numbers into the equation:

$$\frac{dx}{dt} = 2 \times 10^{-3}$$

"That monster!" cried Recordis, who had figured out that the sand would indeed run dry after 48 hours.

**3.** "We'll integrate to find the answer," the king said.

$$\frac{dn}{dt} = 181e^{-2t} - 1$$

$$n = \int(181e^{-2t} - 1) \, dt$$

We used the sum rule and the multiplication rule:

$$n = 181 \int e^{-2t} \, dt - \int dt$$

$$= 181 \int e^{-2t} \, dt - t$$

Trigonometeris suggested that we try the substitution $u = -2t$; $dt = -\frac{1}{2}du$.

$$n = 181 \int e^u \left(-\frac{1}{2}\right) du - t$$

We recognized $e^u$ as the indestructible function whose derivative and integral both equal itself:

$$n = -90.5e^u - t + C$$

We made the reverse substitution, $u = -2t$:

$$n = 90.5e^{-2t} - t + C$$

Next, we solved for the arbitrary constant $C$ by using the initial condition $n = 0$ when $t = 0$ (because, unfortunately, we had solved zero problems when the gremlin first gave us the test).

$$0 = -90.5e^0 - 0 + C$$

$$C = 90.5$$

The final equation for $n$ became:

$$n = -90.5e^{-2t} - t + 90.5$$

We found the value of $n$ when $t = 48$:

$$n = -90.5e^{-96} - 48 + 90.5$$

$$= (-90.5)(2 \times 10^{-42}) + 42.5$$

"If you put that in round numbers, $10^{-42}$ is practically zero," the professor noted. "That means $n \simeq 43$."

"That villain!" Recordis cried. "We need to do 45 problems! If he's right, we'll never get finished in time. I hope he's wrong."

**4.** Polynomials can be differentiated using the sum rule and the power rule for derivatives:

$$f'(x) = 3x^2 - 6x + 4$$

**5.** $f'(x) = 96x^{95} + 430x^{42} - 60x^{19} + 11$

**6.** We wrote this expression as a polynomial with negative exponents:

$$f'(x) = 2x + 1 + 0 - x^{-2} - 2x^{-3} - 3x^{-4}$$

**7.** The professor wrote down the chain rule as it applied to a function raised to a power:

$$f(x) = u^n$$

$$f'(x) = nu^{n-1} \frac{du}{dx}$$

We wrote our function like this:

$$u = 4 - x^3$$

$$f(u) = u^{1/2}$$

$$f'(x) = \tfrac{1}{2}(4 - x^3)^{-1/2}(-3x^2) = \frac{-3}{2}(x^2)(4 - x^3)^{-1/2}$$

**8.** We wrote this function as a product and used the product rule:

$$f(x) = (x^4 + 10)(3 - x^2)^{-1}$$

$$f'(x) = (x^4 + 10) \frac{d}{dx}(3 - x^2)^{-1} + (3 - x^2)^{-1} \frac{d}{dx}(x^4 + 10)$$

$$= \frac{2x(x^4 + 10)}{(3 - x^2)^2} + \frac{4x^3}{3 - x^2}$$

**9.** Let $u = (x + 4)^3$. Also,

$$f(u) = (3 + u)^2$$

$$\frac{du}{dx} = 3(x + 4)^2$$

$$\frac{df}{du} = 2(3 + u)$$

Using the chain rule gives:

$$f'(x) = 6[3 + (x + 4)^3](x + 4)^2$$

**10.** "This calls for the product rule," Recordis said.

$$f'(x) = x^3 \frac{d}{dx}(3 + x^2)^{1/2} + (3 + x^2)^{1/2} \frac{d}{dx} x^3$$

$$= \frac{x^4}{\sqrt{3 + x^2}} + 3x^2 \sqrt{3 + x^2}$$

**11.** Recordis almost fainted until he remembered the method of logarithmic implicit differentiation.

$$y = \sqrt{\frac{(x + 5)(x - 3)}{(x + 2)(x - 1)}}$$

$$\ln y = \tfrac{1}{2}[\ln(x + 5) + \ln(x - 3) - \ln(x + 2) - \ln(x - 1)]$$

$$\frac{1}{y}\frac{dy}{dx} = \frac{1}{2}\frac{d}{dx}[\ln(x + 5) + \ln(x - 3) - \ln(x + 2) - \ln(x - 1)]$$

$$\frac{dy}{dx} = \frac{y}{2}\left[\frac{1}{x + 5} + \frac{1}{x - 3} - \frac{1}{x + 2} - \frac{1}{x - 1}\right]$$

$$= \frac{1}{2}\sqrt{\frac{(x + 5)(x - 3)}{(x + 2)(x - 1)}}\left[\frac{1}{x + 5} + \frac{1}{x - 3} - \frac{1}{x + 2} - \frac{1}{x - 1}\right]$$

**12.** Without thinking, we used the chain rule and the power rule:

$$f(x) = \sin^2 x + \cos^2 x$$

$$f'(x) = 2 \sin x \cos x - 2 \cos x \sin x$$

$$= 0$$

"Of course it should be zero!" Trigonometeris said, realizing the obvious. "We know that $\sin^2 x + \cos^2 x = 1$ for any value of $x$, so this function is really a constant function. It better have a derivative of zero."

**13.** "This is hopeless!" the king said.

"We need to use the product rule several times," the professor told him.

$$f'(x) = x^2 \frac{d}{dx}(e^x \sin x \ln x) + e^x \sin x \ln x \frac{d}{dx}(x^2)$$

"The last one is easy," Recordis said; "$(d/dx)x^2 = 2x$."
We used the product rule again on the first part:

$$\frac{d}{dx}(e^x \sin x \ln x) = e^x \frac{d}{dx}(\sin x \ln x) + (\sin x \ln x)\frac{d}{dx}e^x$$

"That last one is easy," Recordis said again; "$(d/dx)e^x = e^x$."

$$\frac{d}{dx}(\sin x \ln x) = \sin x \frac{d}{dx}\ln x + \ln x \frac{d}{dx}\sin x$$

"Those two are both easy!" Recordis said.

$$\frac{d}{dx}\ln x = \frac{1}{x}$$

$$\frac{d}{dx}\sin x = \cos x$$

Now all we had to do was put the pieces together:

$$f'(x) = xe^x \sin x + x^2 e^x \ln x \cos x + x^2 e^x \sin x \ln x + 2xe^x \sin x \ln x$$

**14.** $f(x) = e^{\ln x}$

"That's trivial!" Recordis said. "By definition, $e^{\ln x} = x$. That means $f(x) = x$, so $f'(x) = 1$."

**15.** "It's obvious what the approximate shape of our course would be," the king said, "although I do *not* intend to escape anywhere. We would travel straight along the shore for some distance (call it $x$) and then board our boat and sail in a straight line to the island." (Figure 16–5.)

"That means that all we need to do is figure out what $x$ is," Recordis said.

"I know what the total time is," the professor said.

$$\text{(total time)} = T = \frac{\text{(distance on land)}}{\text{(speed on land)}} + \frac{\text{(distance on water)}}{\text{(speed on water)}}$$

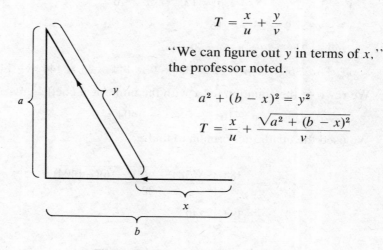

$$T = \frac{x}{u} + \frac{y}{v}$$

"We can figure out $y$ in terms of $x$," the professor noted.

$$a^2 + (b - x)^2 = y^2$$

$$T = \frac{x}{u} + \frac{\sqrt{a^2 + (b - x)^2}}{v}$$

**Figure 16–5.**

"$T$ is the function we want to minimize," the king said. "We now have $T$ expressed as a function of $x$, so all we need to do is find the derivative $dT/dx$ and set it equal to zero."

$$\frac{dT}{dx} = \frac{1}{u} + \frac{1}{v} \tfrac{1}{2}[a^2 + (b - x)^2]^{-1/2} (2)(b - x)(-1)$$

$$= \frac{1}{u} - \frac{b - x}{v\sqrt{a^2 + (b - x)^2}}$$

Setting the derivative equal to zero to find the optimum $x$ gave:

$$\frac{1}{u} = \frac{b - x}{v\sqrt{a^2 + (b - x)^2}}$$

$$\frac{1}{u^2} = \frac{(b - x)^2}{v^2[a^2 + (b - x)^2]}$$

$$v^2[a^2 + (b - x)^2] = u^2(b - x)^2$$

$$v^2a^2 + v^2(b^2 - 2bx + x^2) = u^2(b^2 - 2bx + x^2)$$

$$v^2a^2 + v^2b^2 - 2bxv^2 + v^2x^2 = u^2b^2 - 2bxu^2 + u^2x^2$$

$$x^2(v^2 - u^2) + x(2bu^2 - 2bv^2) + (v^2a^2 + v^2b^2 - u^2b^2) = 0$$

"That looks pretty hopeless," the professor said.

"We know the numbers," Recordis pointed out. "It will help if we put those numbers in the expressions for the coefficients of $x$."

The gremlin had given us these numbers:

$$v = 3, \quad u = 5, \quad b = 12, \quad a = 10$$

$$v^2 - u^2 = 9 - 25 = -16$$

$$2bu^2 - 2bv^2 = (24)(16) = 384$$

$$v^2a^2 + v^2b^2 - u^2b^2 = 9 \times 100 + 9 \times 144 - 25 \times 144 = -1404$$

We rewrote the equation for $x$ with the numerical coefficients:

$$-16x^2 + 384x - 1404 = 0$$

We used the quadratic formula to find $x$:

$$x = \frac{-384 \pm \sqrt{384^2 - (4)(-16)(-1404)}}{-32}$$

$$= \frac{-384 \pm 240}{-32}$$

$$= 19.5 \quad \text{or} \quad = 4.5$$

"That first answer must obviously be an extraneous root," the king pointed out. "It's clear from the map that $x$ can't be greater than 12."

"Then the answer must be $x = 4.5$," the professor said.

**16.** The professor decided to use the method of implicit differentiation, so we took $d/dx$ and applied it to both sides of the equation:

$$\frac{d}{dx}(x^2 + y^2) = \frac{d}{dx}r^2$$

$$2x \frac{dx}{dx} + 2y \frac{dy}{dx} = 0$$

$$\frac{dy}{dx} = \frac{-x}{y}$$

**17.** We applied the same method to this problem. (Notice that this is the equation of an ellipse.)

$$\frac{d}{dx}\left(\frac{x^2}{a^2} + \frac{y^2}{b^2}\right) = 0$$

$$\frac{x}{a^2} + \frac{y}{b^2} \frac{dy}{dx} = 0$$

$$\frac{dy}{dx} = -\frac{b^2 x}{a^2 y}$$

**18.** First we simplified the expression:

$$(x + y)^{1/2}(x - y)^{1/2} = 1$$

$$(x^2 - xy + xy - y^2)^{1/2} = 1$$

$$(x^2 - y^2)^{1/2} = 1$$

$$x^2 - y^2 = 1$$

"There are two choices now," the professor said. "We can solve for $y$ first, or we can use the implicit differentiation method." The king and Recordis disagreed about which method would be quicker, so they had a race.

| Implicit Method (King) | Explicit Method (Recordis) |
|---|---|
| $\dfrac{d(x^2 - y^2)}{dx} = \dfrac{d}{dx} 1$ | $y^2 = x^2 - 1$ |
| $2x - 2y \dfrac{dy}{dx} = 0$ | $y = \sqrt{x^2 - 1}$ |
| $2y \dfrac{dy}{dx} = 2x$ | $\dfrac{dy}{dx} = \frac{1}{2}(x^2 - 1)^{-1/2}(2x)$ |
| $\dfrac{dy}{dx} = \dfrac{x}{y}$ | $\dfrac{dy}{dx} = \dfrac{x}{\sqrt{x^2 - 1}}$ |

"We got different answers!" Recordis exclaimed.

"No, you did not get different answers," the professor said. "If you make the substitution $y = \sqrt{x^2 - 1}$, you will see that you have exactly the same answer that the king has."

**19.** Using the implicit method again gave us:

$$\frac{d}{dx}(x-3)y^2 = \frac{d}{dx}(x^2+4)$$

$$(x-3)\frac{d}{dx}y^2 + y^2\frac{d}{dx}(x-3) = 2x$$

$$(x-3)2y\frac{dy}{dx} + y^2 = 2x$$

$$\frac{dy}{dx} = \frac{2x - y^2}{2y(x-3)}$$

**20.** "This is a regular integration problem," the professor said. We let $s$ stand for Rutherford's position:

$$\frac{ds}{dt} = 3 - t^2; \qquad s = 0 \text{ when } t = 0$$

$$s = \int(3 - t^2)\,dt$$

$$= 3t - \frac{t^3}{3} + C$$

Solving for $C$, we had:

$$0 = C$$

$$s = 3t - \frac{t^3}{3}$$

Now we needed to find out what $t$ equaled when Rutherford came to a stop. That meant we had to find $t$ when $ds/dt = 0$:

$$0 = 3 - t_{stop}^2$$

$$t_{stop} = \sqrt{3}$$

We put that result back into the equation for $s$:

$$s = 3\sqrt{3} - \frac{3\sqrt{3}}{3} = 2\sqrt{3} = 3.46$$

**21.** We set up an integral:

$$\frac{dv}{v} = -2dt$$

$$\int v^{-1}\,dv = -2t$$

$$\ln v = -2t + C$$

We used the initial condition:

$$\ln 5 = C$$

$$\ln v = -2t + \ln 5$$

The professor suggested we take both sides of the last equation and raise $e$ to that power:

$$e^{\ln v} = e^{-2t}\, e^{\ln 5}$$

$$v = 5e^{-2t}$$

Now we had to solve for the value of $t$ that made $v = 0.01$:

$$0.01 = 5e^{-2t}$$

$$t = \tfrac{1}{2} \ln 500$$

$$= 3.1$$

"Now we have to integrate the equation $ds/dt = 5e^{-2t}$ to find the position," Recordis said.

$$s = \frac{-5}{2}e^{-2t} + C$$

Using the initial condition $s = 0$ when $t = 0$ gave us:

$$C = \frac{5}{2}$$

$$s = \frac{-5}{2}e^{-2t} + \frac{5}{2}$$

"Now we just about have it," the king said. "All we need to do is put the value $t = 3.1$ into the equation for $s$, and then we will know how far Mongol travels before his speed equals 0.01."

$$s = \frac{-5}{2}e^{-6.2} + \frac{5}{2}$$

$$= 2.49$$

**22.** By this time Recordis was an expert at integrating polynomials (since they are his favorite kind of function):

$$x^4 - x^2 + 5x + C$$

**23.** The professor suggested the substitution:

$$u = x^2 - 1$$

$$\frac{du}{dx} = 2x$$

$$\int x\sqrt{x^2 - 1}\, dx = \tfrac{1}{2} \int u^{1/2}\, du = \tfrac{1}{3}u^{3/2} + C$$

$$= \tfrac{1}{3}(x^2 - 1)^{3/2} + C$$

**24.** ''We already did this one!'' Recordis said gleefully. ''We can just look it up in our table.''

$$\int \frac{1}{1 + x^2} \, dx = \arctan x + C$$

**25.** We used Trigonometeris' identity for $\sin^2 x$ to simplify matters a little bit:

$$I = \int 4x^2 \sin^2 x \, dx = \int 4x^2(\tfrac{1}{2})(1 - \cos 2x) \, dx$$

$$= 2 \int x^2 \, dx - 2 \int x^2 \cos 2x \, dx$$

$$= \tfrac{2}{3}x^3 - 2 \int x^2 \cos 2x \, dx$$

The professor suggested the substitution $z = 2x$:

$$I = \tfrac{2}{3}x^3 - \tfrac{1}{4} \int z^2 \cos z \, dz$$

''We'll have to use integration by parts,'' the professor said. We made a hazardous guess at the parts and hoped it would work:

Let $u = z^2$. Also,

$$dv = \cos z \, dz$$

$$du = 2z \, dz$$

$$v = \sin z$$

Using the formula for integration by parts gave us:

$$I_2 = \int z^2 \cos z \, dz$$

$$= uv - \int v \, du$$

$$= z^2 \sin z - \int \sin z \, (2z) \, dz$$

''The new integral is simpler,'' the professor said encouragingly. ''Maybe if we try integration by parts once more, we'll be able to get it.''

$$I_3 = \int z \sin z \, dz$$

Let $u = z$. Also,

$$dv = \sin z \, dz$$

$$du = dz$$

$$v = -\cos z$$

$$I_3 = uv - \int v \, du$$

$$= -z \cos z + \int \cos z \, dz$$

"That integral is easy," Recordis said.

$$I_3 = -z \cos z + \sin z + C$$

Now all we had to do was put all the other subintegrals back together and make the reverse substitution, $z = 2x$:

$$\int 4x^2 \sin^2 x \, dx = \tfrac{2}{3}x^3 - x^2 \sin 2x - x \cos 2x + \tfrac{1}{2} \sin 2x + C$$

**26.** $dy/dx = 5x + 4$. Also,

$$y = \int(5x + 4) \, dx$$

$$= \frac{5}{2}x^2 + 4x + C$$

$$C = -15$$

$$y = \frac{5}{2}x^2 + 4x - 15$$

**27.** "We should be able to integrate this equation twice," the professor said. We let $v = dy/dt$, so

$$\frac{dv}{dt} = 5$$

$$v = 5t + C$$

"We have an initial condition for $dy/dt$," Recordis noted.
"That's lucky," the professor said.

$$\frac{dy}{dt} = 5t + 3$$

$$y = \frac{5}{2}t^2 + 3t + C$$

Using the other initial condition, we got:

$$C = 2$$

The final answer became:

$$y = \frac{5}{2}t^2 + 3t + 2$$

**28.** We did this problem in the same way as the preceding one:

$$v = t^2 + 0$$

$$y = \tfrac{1}{3}t^3$$

**29.** Two forces were acting on Mongol, so the total force was

$$F = F_{\text{spring}} + F_{\text{friction}} = -500x - 10 \frac{dx}{dt}$$

We used the equation that Builder gave us:

$$F = m \frac{d^2x}{dt^2}$$

$$-500x - 10 \frac{dx}{dt} = m \frac{d^2x}{dt^2}$$

Using the value $m = 1000$, we rewrote this equation as a second-order linear homogeneous constant-coefficient differential equation:

$$\frac{d^2x}{dt^2} + \frac{1}{100} \frac{dx}{dt} + \tfrac{1}{2}x = 0$$

We set up the characteristic equation:

$$r^2 + \frac{1}{100}r + \tfrac{1}{2} = 0$$

$$r = \frac{-0.01 \pm i\, 1.414}{2}$$

We set up the solution:

$$x = e^{-0.005t}\,(A \sin 0.707t + B \cos 0.707t)$$

Using the initial conditions:

$$20 = B, \qquad 0.14 = A$$

we obtained the final solution:

$$x = e^{-0.005t}(0.14 \sin 0.707t + 20 \cos 0.707t)$$

**30.** The professor was bothered by what a unit of work was, but we realized we didn't have time to worry about that now. The integral was straightforward, with the result (work) $= 9k/100$.

**31.** We set up the partial fractions:

$$\frac{x-1}{(x-2)(x-3)} = \frac{A}{x-2} + \frac{B}{x-3}$$

$$x - 1 = A(x-3) + B(x-2)$$

coefficients of $x$:

$$1 = A + B$$

constant terms:

$$-1 = -3A - 2B$$

The result was that $A = -1$, $B = 2$.

$$\frac{x - 1}{(x - 2)(x - 3)} = \frac{2}{x - 3} - \frac{1}{x - 2}$$

$$\int_4^5 \frac{x - 1}{(x - 2)(x - 3)} \, dx = 2 \ln(x - 3) \Big|_4^5 - \ln(x - 2) \Big|_4^5$$

$$= 2 \ln 2 - 2 \ln 1 - \ln 3 + \ln 2$$

$$= 3 \ln 2 - \ln 3$$

$$\int_4^5 \frac{x - 1}{(x - 2)(x - 3)} \, dx = 0.98$$

**32.** This was easy:

$$(4)(\tfrac{1}{3}) \, x^3 \Big|_0^{10} = 1333$$

**33.** The king suggested the substitution $u = 2x$.

$$\int_0^{\pi/6} \tan 2x \, dx = \frac{1}{2} \int_0^{\pi/3} \tan u \, du$$

$$= -\tfrac{1}{2} \ln \cos u \Big|_0^{\pi/3}$$

$$= -\tfrac{1}{2} \ln \cos\left(\frac{\pi}{3}\right) + \tfrac{1}{2} \ln \cos 0$$

$$= -\tfrac{1}{2} \ln \tfrac{1}{2}$$

$$\int_0^{\pi/6} \tan 2x \, dx = 0.35$$

**34.** $e^x \Big|_0^{10} - \tfrac{1}{4} x^4 \Big|_0^{10} = 22{,}026 - 1 - 2500 = 19{,}525$

**35.** This looked hard until the king remembered a perfect substitution:

Let $u = x^5$, $\quad du = 5x^4 \, dx$.

$$\int_1^2 x^4 \sin(x^5) \, dx = \frac{1}{5} \int_1^{32} \sin u \, du = \frac{-1}{5} (\cos 32 - \cos 1)$$

$$= 0.059$$

**36.** Just as we were ready to try integration by parts, Trigonometeris remembered a formula to make matters much simpler:

$$\sin x \cos x = \tfrac{1}{2} \sin 2x$$

$$\int_0^{\pi/2} \sin x \cos x \, dx = \frac{1}{2} \int_0^{\pi/2} \sin 2x \, dx = -\tfrac{1}{4} \cos 2x \Big|_0^{\pi/2}$$

$$= -\tfrac{1}{4} (\cos \pi - \cos 0) = \tfrac{1}{2}$$

**37.** We realized that we had no choice but integration by parts. We decided to treat the integral as if it were an indefinite integral first, and then use the limits of integration after we had found the antiderivative function:
Let $u = x^2$. Also,

$$dv = \sin x \, dx$$

$$du = 2x \, dx$$

$$v = -\cos x$$

$$\int x^2 \sin x \, dx = -x^2 \cos x + \int \cos x \, 2x \, dx$$

We used integration by parts again:
Let $u = x$. Also,

$$du = dx$$

$$dv = \cos x \, dx$$

$$v = \sin x$$

$$\int x^2 \sin x \, dx = -x^2 \cos x + 2x \sin x - \int 2\sin x \, dx$$

$$\int_2^3 x^2 \sin x \, dx = (-x^2 \cos x + 2x \sin x + 2\cos x)\Big|_2^3$$

$$= -7 \cos 3 + 6 \sin 3 + 2 \cos 2 - 4 \sin 2$$

$$= 3.3$$

**38.** We defined a new function equal to the distance between the two curves: $y = y_1 - y_2 = \sin x - x^2 + \pi x$.
"Now we need to know what limits of integration to use," Recordis said.

After making a quick sketch of both curves, it was clear that we wanted to integrate between the two points where the curves intersected. One intersection point came at $x = 0$, because then both $y_1$ and $y_2$ were equal to zero. We found that the other intersection point occurred where $x = \pi$, so we integrated from $x = 0$ to $x = \pi$ to find the total area between the curves:

$$A = \int_0^\pi (\sin x - x^2 + \pi x) \, dx = (-\cos x - \tfrac{1}{3}x^3 + \tfrac{1}{2}\pi x^2)\Big|_0^\pi$$

$$= 1 + 1 - \tfrac{1}{3}\pi^3 + \tfrac{1}{2}\pi^3 = 7.17$$

**39.** Realizing that the balancing point must lie along the central line, we used Builder's formula for the center of mass:

$$x_{\text{com}} = \frac{\int x \, dm}{\rho h a^2}$$

Then we calculated what the differential mass element $dm$ would be (Figure 16–6):

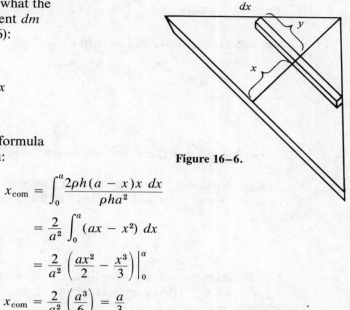

$$dm = \rho \, dV$$
$$= 2\rho hy \, dx$$
$$y = a - x$$

We put these into the formula and did the integration:

**Figure 16–6.**

$$x_{com} = \int_0^a \frac{2\rho h(a - x)x \, dx}{\rho ha^2}$$

$$= \frac{2}{a^2} \int_0^a (ax - x^2) \, dx$$

$$= \frac{2}{a^2} \left( \frac{ax^2}{2} - \frac{x^3}{3} \right) \Bigg|_0^a$$

$$x_{com} = \frac{2}{a^2} \left( \frac{a^3}{6} \right) = \frac{a}{3}$$

**40.** We divided the football into a collection of little cylinders:

$$dV = \pi y^2 \, dx$$
$$y = \sin x$$

Now all we had to do was integrate from $x = 0$ to $x = \pi$:

$$V = \pi \int_0^\pi \sin^2 x \, dx$$

$$= \pi \int_0^\pi \tfrac{1}{2}(1 - \cos 2x) \, dx$$

$$= \tfrac{1}{2}\pi^2$$

**41.** We found the first derivative to look for horizontal tangents:

$$y = e^{-x^2}$$

$$\frac{dy}{dx} = -2xe^{-x^2}$$

"Where does that equal zero?" Recordis asked.

"The only point is where $x = 0$," the professor said. "That means that this curve has only one horizontal tangent."

We calculated the second derivative:

$$\frac{d^2y}{dx^2} = (4x^2 - 2)(e^{-x^2})$$

The second derivative will be zero if

$$4x^2 - 2 = 0$$

$$x^2 = \tfrac{1}{2}$$

$$x = \pm \frac{1}{\sqrt{2}}$$

"That means that there are two points of inflection," the professor noted.

"One more thing," the king said. "We need to check whether the second derivative is positive or negative at the point where the horizontal tangent occurs."

$$x = 0, \qquad \frac{d^2y}{dx^2} = -2$$

"That means the point is a maximum," the king said.

"I remember," Trigonometeris stated. "If the second derivative is negative, the curve spills water."

We could easily see that $y = 1$ when $x = 0$. We could also see that the function was always greater than 0 but less than 1, so Igor was able to sketch the graph (Figure 16–7).

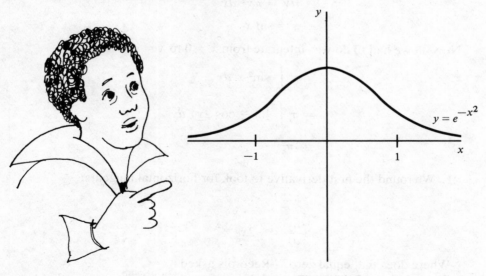

**Figure 16–7.**

42. First we found the homogeneous solution:

$$\frac{d^2y}{dx^2} + 2\,\frac{dy}{dx} + 2y = 0$$

We set up the characteristic equation:

$$r^2 + 2r + 2 = 0$$

$$r = \frac{-2 \pm \sqrt{4 - 8}}{2}$$

$$= -1 + i, \qquad = -1 - i$$

Since we had a complex answer, we set up an exponential and a trigonometric solution:

$$y = e^{-x}(A \cos x + B \sin x)$$

Now we needed to guess a particular solution to the nonhomogeneous equation. We spent hours trying to find a solution that would work. After each guess that failed we came up with a new guess that was even more complicated than the preceding one. No guess, no matter how complicated, worked.

"If only we just had $2y = 10$," Recordis moaned. "Then we could say $y = 5$. If only we knew a function that would allow us to ignore those derivatives."

"That wouldn't help," the professor said. "We know that the only kind of function that has a derivative of zero is a constant function."

"That's it!" the king exclaimed. "We'll use a constant function! Let's guess $y = 5$."

"We can't use a constant function!" Recordis said. "This is a differential equation. Differential equations are hard, and they always have complicated solutions."

We tried the king's guess anyway, and found that it did work:

$$\frac{d^2}{dx^2} 5 + \frac{d}{dx} 5 + 2 \times 5 = 10$$

"That is the particular solution we need," the professor said. We found the complete solution by adding the particular solution to the homogeneous solution:

$$y = 5 + e^{-x}(A \cos x + B \sin x)$$

**43.** We realized that we would have to use the curve length integral. First we found $dy/dx$:

$$\frac{dy}{dx} = \tfrac{1}{2}(e^{x/36} - e^{-x/36})$$

$$\left(\frac{dy}{dx}\right)^2 = \tfrac{1}{4}(e^{x/18} - 2 + e^{-x/18})$$

$$1 + \left(\frac{dy}{dx}\right)^2 = \tfrac{1}{4}(e^{x/18} + 2 + e^{-x/18})$$

$$\sqrt{1 + (dy/dx)^2} = \tfrac{1}{2}(e^{x/36} + e^{-x/36})$$

Now we set up the length integral:

$$L = \frac{1}{2} \int_{-50}^{50} (e^{x/36} + e^{-x/36})\, dx$$
$$= 18(e^{x/36} - e^{-x/36})\Big|_{-50}^{50}$$
$$= 135.4$$

**44.** We set up the function $y = y_2 - y_1 = x - x^2$, and integrated it from $x = 0$ to $x = 1$ in order to find the total area between the two curves:

$$A = \int_0^1 (x - x^2)\, dx = \tfrac{1}{2}x^2 - \tfrac{1}{3}x^3 \Big|_0^1 = \tfrac{1}{2} - \tfrac{1}{3} = \tfrac{1}{6}$$

**45.** Our sand was just about ready to run out. We had finished all the problems except this one.

"We've got to hurry!" Recordis said. "We have only that much sand left!" He held his fingers a tiny distance apart.

The professor desperately guided Igor through the required calculations.

"It can't be done," Recordis moaned. "There is no answer!"

"That's the answer!" the professor cried.

"What's the answer?"

"There is no answer! Look at the graph (Figure 16–8). There is no way to find a derivative for the absolute-value function at that point, because there is no real tangent line. In fact, I bet that's true for any function with a cusp in it—you would have to say that the function is nondifferentiable at the point where the cusp is. So the answer must be that there is no answer! The gremlin threw an impossible problem into the test to confuse us."

"That gremlin!" Recordis cried. "I hate him!" He whipped out his pen and wrote "no answer" on the large scroll where he had been keeping track of the answers to the test.

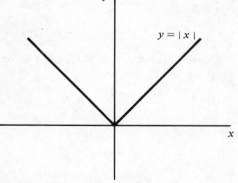

**Figure 16–8.**

At the instant he wrote the last letter, the very last grain of sand hesitated for a moment at the vertex of the hourglass cone, and then dropped to the pile of sand below. There was a trumpet call from the hallway and in marched the gremlin, wearing an impressive royal robe and followed by two short gremlin attendants, each carrying a trumpet.

"I want you to know that I inspected the palace and I like it very much, although there are a few parts that will have to be changed to suit my taste. It's been nice knowing you. I hope you are all packed." We all looked at him in terror. "Well, which one did you in? If you really want to give me the satisfaction, I will look over your answers and see where you went wrong."

"I think we got them all," the professor said timidly.

"Are you sure?" Recordis asked. "We were in such a hurry that we may have misplaced one or two problems."

"We did get them all, you vile creature," the king said, standing straight up and staring the gremlin in the eye. He took the answer scroll from Recordis and handed it to the gremlin. He remained standing there while the gremlin slowly read through the entire scroll. The next few minutes seemed to take forever, but the king did not budge. A look of consternation grew on the gremlin's face as he read the answer to each problem that he was sure would have ruined us.

"You can't have solved the last one!" he said desperately, but when he saw Recordis' writing he fell to the floor sobbing. "Curses! I'm ruined!" he moaned. "All that effort wasted! All is lost! You know everything worth knowing about single-variable calculus!"

A large puddle of tears developed around the prostrate gremlin. Recordis felt so sorry for him that he patted him on the back. "It can't be that bad," he said consolingly. "I'm sure you'll find other complications. You'll think of something like multivariable calculus or scientifically applied problems or . . ."

"Shut up, Recordis!" the professor cried. "What are you saying?"

"Enough of that!" the gremlin said, suddenly standing up. "I shouldn't have lost control of myself like that. I can make things more complicated, and I will. It is just a matter of time before you discover another subject, and then I shall win and rule Carmorra!" He grabbed his two attendants and whooshed out the window.

"We surely got rid of him," Recordis said, and we all laughed, for the first time in a long while.

The party the next day was very festive and bright. The shining ribbons decorated the grounds. The roses in the garden were all in bloom, and Spike Rock was safely covered. The children loved the doughnuts, the ice cream cones, and the ride on the spring.

That night the king asked me how long I could stay. "I will have to return home when I regain my memory," I said, "but I will be glad to stay in Carmorra for the present."

"We deeply appreciate your services," the king said. "If you ever want to return home, we'll do our best to help you, even if we need to invent a new subject."

As we looked out at the moon, Farmer Floran said, "I wonder whether you could figure out how the moon moves."

"We wouldn't have had a chance before," the king said. "Maybe we could now."

This brings to an end my part of the story. There were indeed more adventures, and we were constantly amazed at the applications of the subject of calculus. I'll let the professor summarize what we did in her book, which she graciously granted me permission to reprint here as Chapter 17. "It's been great having another woman in the Main Conference Room," she said wistfully.

I hope that this entire account has been as much fun for the reader as it was for me, and that it will be beneficial to anyone who is interested in discovering the mysteries of calculus.

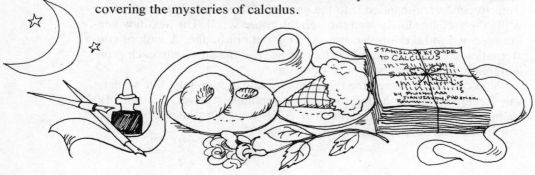

# 17
# Stanislavsky Guide to Calculus:
## The Complete, Authoritative, Summary Guide to the Mysterious Subject of Calculus

*by Professor A. A. A. Stanislavsky, Ph.D., etc., etc.*
*Royal Institute of Carmorra*

Calculus begins by trying to solve the following problem: What is the slope of the tangent line for a particular curve at a given point? It turns out that this problem is exactly the same problem as finding the speed of an object if we are given its position function.

The slope of the tangent line to the curve $y = f(x)$ is given by the following expression, known as the derivative:

$$(\text{slope of tangent line}) = (\text{derivative}) = f'(x) = \frac{dy}{dx}$$

$$= \lim_{\Delta x \to 0} \frac{f(x + \Delta x) - f(x)}{\Delta x}$$

The operation ''lim'' means to take the limit of the expression as $\Delta x$ moves very close to zero, but we don't ever let $\Delta x$ actually equal zero.

We can derive a set of rules that make it possible to find the derivatives of different functions. (We call this process differentiating the function.)

$$y = c \qquad y' = 0$$
$$y = cx \qquad y' = c$$

239

---

## SUM RULE

$$y = f(x) + g(x) \qquad y' = f'(x) + g'(x)$$

---

## PRODUCT RULE

$$y = f(x)\, g(x) \qquad y' = f(x)\, g'(x) + f'(x)\, g(x)$$

---

## POWER RULE

$$y = x^n \qquad y' = nx^{n-1}$$

---

## CHAIN RULE

$$y = g(f(x)) \qquad y' = \frac{dg}{df}\,\frac{df}{dx}$$

---

## TRIGONOMETRIC FUNCTIONS

$$y = \sin x \qquad y' = \cos x$$
$$y = \cos x \qquad y' = -\sin x$$
$$y = \tan x \qquad y' = \sec^2 x$$
$$y = \text{ctn } x \qquad y' = -\csc^2 x$$
$$y = \sec x \qquad y' = \sec x \tan x$$
$$y = \csc x \qquad y' = -\csc x \text{ ctn } x$$
$$y = \arcsin x \qquad y' = (1 - x^2)^{-1/2}$$
$$y = \arctan x \qquad y' = (1 + x^2)^{-1}$$

---

## EXPONENTIAL FUNCTIONS

$$y = a^x \qquad y' = (\ln a)a^x$$

---

## LOGARITHM FUNCTIONS

$$y = \ln x \qquad y' = \frac{1}{x}$$

It is often useful to reverse the process of differentiation. We call this process *integration*. We make this definition for the indefinite integral:

$$\int f(x)\ dx = F(x) + C \qquad \text{if and only if } \frac{dF}{dx} = f(x)$$

We can then make a set of rules to calculate integrals:

---

## PERFECT INTEGRAL RULE

$$\int dx = x + C$$

---

## SUM RULE

$$\int (f(x) + g(x))\ dx = \int f(x)\ dx + \int g(x)\ dx$$

---

## PRODUCT RULE

$$\int af(x)\ dx = a \int f(x)\ dx \qquad \text{if } a \text{ is constant}$$

---

## POWER RULE

$$\int x^n\ dx = \frac{1}{n+1}\ x^{n+1} + C \qquad \text{if } n \neq -1$$

$$\int x^{-1}\ dx = \ln|x| + C$$

---

## TRIGONOMETRIC INTEGRALS

$$\int \sin x\ dx = -\cos x + C$$
$$\int \cos x\ dx = \sin x + C$$
$$\int \tan x\ dx = \ln|\sec x| + C$$
$$\int \sec x\ dx = \ln|\sec x + \tan x| + C$$

---

An integral that contains the quotient of two polynomials can be solved by the method of partial fractions. An integral that contains an expression such as $\sqrt{1 + x^2}$ or $(1 - x^2)^{-1}$ can be solved by using a trigonometric substitution, based on the identity

$$\sin^2 \theta + \cos^2 \theta = 1 \qquad \text{or} \qquad 1 + \tan^2 \theta = \sec^2 \theta$$

An integral containing the product of two different types of functions, or some other hard integrals, can be solved by using the method of integration by parts.

## INTEGRATION BY PARTS

$$\int u \, dv = uv - \int v \, du$$

The amazing feature of integral calculus is that an integral represents not only the antiderivative but also the area under a curve. This is often written as a definite integral.

$$(\text{area under } f(x), \text{ from } x = a \text{ to } x = b) = \int_a^b f(x) \, dx$$

$$= F(b) - F(a)$$

where $F(x)$ is a function such that $dF/dx = f(x)$.

Integrals can be used for more general problems, such as volumes, surface areas, or the center of mass, by comparing the formula for an integral with the formula for a continuous sum:

$$\int_a^b f(x) \, dx = \lim_{\Delta x \to 0} \sum_{i=1}^{n} f(x_i) \, \Delta x \qquad (x_1 = a, \quad x_n = b)$$

The natural logarithm function is defined in terms of definite integrals:

$$\ln x = \int_1^x \frac{1}{t} \, dt$$

The base of the natural logarithm function is the mysterious number $e$, approximately equal to 2.71:

$$e^{\ln x} = x$$

# Appendix
# Answers
# to
# Exercises

**1.** and **2.** See Table A–1.

**Table A–1**

| Points | Slope of Secant Line |
|---|---|
| (2, 4), (3, 9) | 5.0 |
| (2, 4), (2.5, 6.25) | 4.5 |
| (2, 4), (2.3, 5.29) | 4.3 |
| (2, 4), (2.1, 4.41) | 4.1 |
| (2, 4), (2.05, 4.2) | 4.0 |
| (1, 1), (2, 4) | 3.0 |
| (1.5, 2.25), (2, 4) | 3.5 |
| (1.7, 2.89), (2, 4) | 3.7 |
| (1.8, 3.24), (2, 4) | 3.8 |
| (1.9, 3.61), (2, 4) | 3.9 |
| (1.95, 3.80), (2, 4) | 4.0 |

**3.** (Slope) $= 2 \cdot 2 = 4$. Point-slope formula: $(y - 4)/(x - 2) = 4$. $y = 4x - 4$.

**4.** (Slope) $= 2 \cdot 7 = 14$. $y = 14x - 49$. When $y = 50$, the $x$ coordinate of the tangent line is $x = 99/14 = 7.0714$. (This is reasonably close to the best decimal approximation, which is $\sqrt{50} \approx 7.0711$.)

**5.** $f(0)$ is undefined, since at $x = 0$ the expression for $f(x)$ is $0/0$. On the graph this is signified by drawing an open circle at the point $x = 0$. It is clear from the graph that $\lim_{x \to 0} f(x) = 8$. (See Figure A–1.)

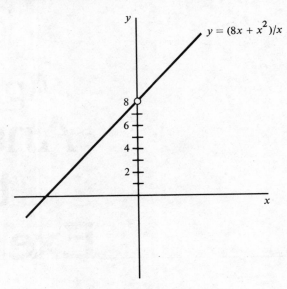

**Figure A–1.**

# Chapter 2

**1.** $y' = 9x^2 + 4x + 1$.

**2.** $y' = 20x^4 + 2x$.

**3.** $y' = a$.

**4.** $y' = 35x^{34}$.

**5.** $y' = x^2 + x + 1$.

**6.** $y' = 8.68x + 0.98$.

**7.** $f(t) = 6t^2 - 15t + 8t - 20$. $f'(t) = 12t - 7$.

**8.** $f(x + \Delta x) = ax^2 + 2ax\,\Delta x + a\,\Delta x^2 + bx + b\,\Delta x + c$.
$f(x + \Delta x) - f(x) = 2ax\,\Delta x + a\,\Delta x^2 + b\,\Delta x$.
$$f'(x) = \lim_{\Delta x \to 0} \frac{2ax\,\Delta x + a\,\Delta x^2 + b\,\Delta x}{\Delta x} = \lim_{\Delta x \to 0} (2ax + a\,\Delta x + b)$$
$$= 2ax + b.$$

**9.** $y + \Delta y = f(x + \Delta x) + g(x + \Delta x) + h(x + \Delta x)$.

$$\frac{\Delta y}{\Delta x} = \frac{f(x + \Delta x) - f(x) + g(x + \Delta x) - g(x) + h(x + \Delta x) - h(x)}{\Delta x}.$$

$$\frac{dy}{dx} = \lim_{\Delta x \to 0} \left[ \frac{f(x + \Delta x) - f(x)}{\Delta x} + \frac{g(x + \Delta x) - g(x)}{\Delta x} \right.$$

$$\left. + \frac{h(x + \Delta x) - h(x)}{\Delta x} \right]$$

$$= f'(x) + g'(x) + h'(x).$$

**10.** (a) $v(t) = dh/dt = -gt + v_0$. (b) $v(0) = v_0$. (c) $dh/dt = 0$ when $0 = -gt + v_0$. $t = v_0/g$.

**11.** (a) $v(t) = -gt$. The velocity is negative, which means that $h$ is becoming smaller (the ball is going down). (b) $h = 0$ when $0 = 64 - \frac{1}{2}gt^2$. $t = 8\sqrt{2}/\sqrt{g}$. (c) When $t = 8\sqrt{2}/\sqrt{g}$, $v(t) = -8\sqrt{2g}$. (d) $3.61 = 8\sqrt{2}/\sqrt{g}$. $g = 9.8$.

**12.** $y' = x^2 - 2x + 3$. Now set $y' = 3$ and solve for $x$: $3 = x^2 - 2x + 3$. $x^2 - 2x = 0$. $x = 0$ or $x = 2$.

**13.** The slope of the curve is $y' = 3x^2$. The slope of the line is $y' = 8$. Set those two slopes equal, and solve for $x$. $x = \pm \sqrt{8/3}$. The line $y = 8x + b$ will be tangent to the curve if it passes through the point $((8/3)^{1/2}, (8/3)^{3/2})$ or the point $(-(8/3)^{1/2}, -(8/3)^{3/2})$. This will happen if $b = 8.7$ or $b = -8.7$.

**14.** First, we need the slope of the line between $(2, 1)$ and $(6, 9)$ (call the slope $m$). $m = (9 - 1)/(6 - 2) = 2$. Now we need to find the point where the slope of the curve $f(x) = -x^2 + 10x - 15$ equals 2. $f'(x) = -2x + 10$. $f'(x)$ equals 2 when $2 = -2x + 10$. $x = 4$.

**15.** Let $f(x) = 3x^2 + x - 4$, and $g(x) = x - 1$. Then calculate the derivatives: $f'(x) = 6x + 1$. $g'(x) = 1$. Now, use l'Hôpital's rule:

$$\lim_{x \to 1} \frac{f(x)}{g(x)} = \lim_{x \to 1} \frac{f'(x)}{g'(x)} = \lim_{x \to 1} \frac{6x + 1}{1} = 7.$$

**16.** Calculate the derivative: $f'(x) = 3x^2$. The formula now becomes:

$$x_{i+1} = x_i - \frac{x_i^3 - 7}{3x_i^2}.$$

For $x_1 = 2$, $x_2 = 2 - 1/12 = 1.917$. $x_3 = 1.917 - 0.041/11.02$. $x_3 = 1.913$. $x_4 = 1.913 - 0.0076/10.978 = 1.9124$. This result is close to the best decimal approximation for $\sqrt[3]{7}$, which is 1.9129.

---

## Chapter 3

**1.** (Acceleration) $= h'' = -g$. The acceleration is negative.

**2.** $h'' = -g$. Note that the acceleration due to gravity is the same whether Mongol drops the ball or throws it into the air.

**3.** $x'' = 0$. Mongol is traveling with a constant velocity. Any object traveling with a constant velocity has zero acceleration.

**4.** $y' = 2x - 3$. $y' = 0$ when $x = 3/2$. $y'' = 2$. Since $y''$ is positive, the curve has a local minimum at $x = 3/2$. The curve never turns down, so the absolute minimum occurs at $x = 3/2$, $y = -9/4$. Since there are no other points where the curve has horizontal tangents, the maximum value of the function must occur at one of the end points of the interval ($x = 0$ or $x = 5$). Such a case is known as a *corner solution*. In this case the maximum value occurs at $x = 5$, $y = 10$.

**5.** $y' = -4x^3 + 16x$. $y'' = -12x^2 + 16$. $y' = 0$ when $x = 0$, $x = 2$, or $x = -2$. At $x = 0$ the curve has a local minimum. At $x = 2$ and $x = -2$ it has local maxima. It is concave upward when $16 - 12x^2$ is positive, which means that $-2/\sqrt{3} < x < 2/\sqrt{3}$.

**6.** $y' = 2x - x^2$. The curve is rising when $0 < x < 2$. $y'' = -2x + 2$. The curve is concave downward when $x > 1$.

**7.** $d^4y/dx^4 = 0$.

**8.** $d^3y/dx^3 = 6a$.

**9.** $d^ny/dx^n = n!$

## Chapter 4

**1.** $f'(t) = 2(3t + 4) + 3(2t - 5)$.

**2.** $y' = \dfrac{2}{3x - 1} - \dfrac{3(2x + 5)}{(3x - 1)^2}$.

**3.** $y' = \dfrac{10}{4x - 3} - \dfrac{40x}{(4x - 3)^2}$.

**4.** $y' = \dfrac{a}{cx + d} - \dfrac{c(ax + b)}{(cx + d)^2}$.

**5.** $y' = (x^2)(3x^2) + (2x)(x^3) = 3x^4 + 2x^4 = 5x^4$.

**6.** $y' = uv\dfrac{dw}{dx} + w\dfrac{d}{dx}(uv) = uv\dfrac{dw}{dx} + w\left(u\dfrac{dv}{dx} + v\dfrac{du}{dx}\right)$

$\quad = uv\dfrac{dw}{dx} + uw\dfrac{dv}{dx} + wv\dfrac{du}{dx}$.

$y = x^3$, $y' = x \times x \times 1 + x \times x \times 1 + x \times x \times 1 = 3x^2$.

**7.** $y' = u(-1)v^{-2}\dfrac{dv}{dx} + \dfrac{du}{dx}\dfrac{1}{v} = \dfrac{v\ du/dx}{v^2} - \dfrac{u\ dv/dx}{v^2}$.

**8.** $dy/du = (3/2)u^{1/2}$. $du/dx = 2x$. $dy/dx = (3/2)(x^2 + 3)^{1/2}(2x)$ $dy/dx = 3x\sqrt{x^2 + 3}$.

**9.** $dy/du = \frac{1}{2}u^{-1/2}$. $du/dx = 2x$. $dy/dx = x(1 + x^2)^{-1/2}$.

**10.** $dy/du = (3/2)u^{1/2}$. $du/dx = 4$. $dy/dx = 6\sqrt{3 + 4x}$.

**11.** $dy/du = \frac{1}{2}u^{-1/2}$. $du/dx = 2ax + b$. $dy/dx = \frac{1}{2}(ax^2 + bx + c)^{-1/2}(2ax + b)$.

**12.** $dy/dv = \frac{1}{2}v^{-1/2}$. $dv/du = -(1/u^2)$. $du/dx = 2x$.

$$\frac{dy}{dx} = \frac{-x}{(x^2 + 4)^2 \sqrt{1 + 1/(x^2 + 4)}}.$$

**13.** $y + \Delta y = \sqrt{x + \Delta x}$.

$$\frac{\Delta y}{\Delta x} = \frac{\sqrt{x + \Delta x} - \sqrt{x}}{\Delta x}.$$

This expression can be simplified by multiplying the top and bottom by $(\sqrt{x + \Delta x} + \sqrt{x})$.

$$\frac{\Delta y}{\Delta x} = \frac{x + \Delta x - x}{\Delta x(\sqrt{x + \Delta x} + \sqrt{x})} = \frac{1}{\sqrt{x + \Delta x} + \sqrt{x}}.$$

$$\frac{dy}{dx} = \lim_{\Delta x \to 0} \frac{1}{\sqrt{x + \Delta x} + \sqrt{x}} = \frac{1}{2\sqrt{x}}.$$

**14.** $y + \Delta y = 1/(x + \Delta x)$.

$$\frac{\Delta y}{\Delta x} = \frac{\frac{1}{x + \Delta x} - \frac{1}{x}}{\Delta x} = \frac{\frac{x}{(x + \Delta x)x} - \frac{x + \Delta x}{(x + \Delta x)x}}{\Delta x}$$

$$= -\frac{1}{(x + \Delta x)x}.$$

$$\frac{dy}{dx} = \lim_{\Delta x \to 0} -\frac{1}{(x + \Delta x)x} = -\frac{1}{x^2}.$$

**15.** $dy/dx = nu^{n-1}(du/dx)$.

**16.** $y' = (4 + 2x)^{-1/2}$.

**17.** $y' = 9.9(4 + 3x)^{2.3}$.

**18.** $y' = \frac{1}{2}(3x^3 + 2x^2 + x)^{-1/2}(9x^2 + 4x + 1)$.

**19.** $y' = -3x(x^2 - 1)^{-3/2}$.

**20.** $y' = \frac{1}{3}x^{-2/3}$.

**21.** $(2x/3)(x^2 + 4)^{-2/3}$.

**22.** $y' = (-1/2x^2)(1 + 1/x)^{-1/2}$.

**23.** $y' = \frac{1}{2}(x + 1/x)^{-1/2}(1 - 1/x^2)$.

**24.** $y' = \dfrac{-2x}{(x^2 + 4)^2}$.

**25.** $\dfrac{d^2A}{dw^2} = -\dfrac{w^3}{(r^2 - w^2)^{1.5}} - \dfrac{3w}{(r^2 - w^2)^{0.5}}$.

When $w = r/\sqrt{2}$, $d^2A/dw^2 = -4$. Since the second derivative is negative, the curve has a local maximum at this point. Recordis is building the largest possible house.

**26.** (a) $2x + 2y(dy/dx) = 0$. $dy/dx = -x/y$.

(b) $y = (r^2 - x^2)^{1/2}$. $dy/dx = -\frac{1}{2}(r^2 - x^2)^{-1/2}(2x) = -x/y$.

**27.** Using the implicit method:

$$\tfrac{2}{9}(y - 1)\,\frac{dy}{dx} - \tfrac{1}{2}(x - 3) = 0.$$

$$\frac{dy}{dx} = \frac{\tfrac{1}{2}(x - 3)}{\tfrac{2}{9}(y - 1)}.$$

## Chapter 5

**1.** $y' = \cos(x^2)2x$.

**2.** $y' = 2\sin x \cos x$.

**3.** $y' = (-1)\sin^{-2} x \cos x = -\text{ctn } x \csc x$.

**4.** $y' = x \cos x + \sin x$.

**5.** $y' = (\sin x)(-\sin x) + (\cos x)(\cos x) = \cos^2 x - \sin^2 x = \cos 2x$.

**6.** (a) $y = \sin(x^2) \cos x + \sin x \cos(x^2)$.

$y' = \sin(x^2)(-\sin x) + (\cos x)(2x)\cos(x^2) + \sin x(2x)(-\sin x^2)$
$\quad + \cos(x^2) \cos x$
$\quad = (2x + 1)(\cos x \cos x^2 - \sin x \sin x^2)$
$\quad = (2x + 1) \cos(x^2 + x)$.

(b) When the chain rule is used, the answer comes directly: $y' = \cos(x^2 + 1)(2x + 1)$.

**7.** If $A$ is measured in degrees and $\theta$ in radians, then $A = (180\theta/\pi)$. $y = \sin(180\theta/\pi)$. $y' = (180/\pi) \cos(180\theta/\pi) = (180/\pi) \cos A$.

**8.** $y' = \cos x$. The curve will have horizontal tangents when $\cos x = 0$. This happens when $x = n\pi + \pi/2$, where $n$ is any integer. When $n$ is even, there is a maximum (as at $x = \pi/2$). When $n$ is odd, there is a minimum (as at $x = 3\pi/2$). $y'' = -\sin x$. The curve is concave downward from $x = 0$ to $x = \pi$; it is concave upward from $x = \pi$ to $x = 2\pi$; etc.

**9.** $y' = \pi \cos(2\pi t/10)$. $y'' = -(\pi^2/5) \sin(2\pi t/10)$.

**10.** $32° = 0.559$. $x = \pi/6$, $y' = \cos(\pi/6) = 0.866$. Using the point-slope formula for the tangent line gives: $(y - \tfrac{1}{2})/(x - \pi/6) = 0.866$. $y = 0.866x + 0.047$. When $x = 0.559$, $y = 0.531$. The best decimal approximation for sin $32° = 0.5299$.

**11.** $y' = \tfrac{1}{2}(1 + \tan^2 \theta)^{-1/2}(2 \tan \theta \sec^2 \theta) = \dfrac{\tan \theta \sec^2 \theta}{\sec \theta} = \tan \theta \sec \theta$.

Remember that $\sec \theta = (1 + \tan^2 \theta)^{1/2}$.

**12.** $y' = \sec^2 \theta = 1/\cos^2 \theta$; $y'$ will never be zero, so this curve has no horizontal tangents. $y'' = 2 \sec \theta \sec \theta \tan \theta$; $y'' = 2 \sec^2 \theta \tan \theta$. $y''$ will be positive when $\tan \theta$ is positive, so the curve will be concave upward in the interval $\theta = 0$ to $\theta = \pi/2$. It will be concave downward in the interval $\theta = -\pi/2$ to $\theta = 0$.

**13.** See Table A–2.

**Table A–2**

| $x$ | $(\sin x)/x$ |
|-----|--------------|
| 0.785 | 0.9004 |
| 0.5 | 0.9589 |
| 0.3 | 0.9851 |
| 0.1 | 0.9983 |
| 0.05 | 0.9996 |

**14.** $dx/dt = (0.8)(3) \cos(3t)$. $d^2x/dt^2 = -(0.8)(9) \sin(3t) = -9x$. From the equation of motion: $-kx = m\, d^2x/dt^2$. $-kx = m(-9x)$. $k = 9m$. $k = 18$.

---

Chapter 6

**1.** Let $r$ equal the radius of the cone formed by the water. $r = ha/2b$. $V = \frac{1}{3}\pi r^2 h = \pi a^2 h^3/12b^2$. $dV/dh = \pi a^2 h^2/4b^2$. $dV/dt = u$. $dh/dt = 4b^2 u/\pi a^2 h^2 = 0.11$ meter/minute.

**2.** Let $x$ be the distance from the wall to the bottom of the ladder, and let $y$ be the distance up the wall to the top of the ladder. $x^2 + y^2 = L^2$. $dx/dt = u$. $dy/dx = -x/\sqrt{L^2 - x^2}$. $dy/dt = -ux/\sqrt{L^2 - x^2}$. (The result is negative because $y$ is becoming smaller as the ladder slides down.)

**3.** Let $D$ be the distance from the point to the curve:

$$D = \sqrt{(x - 4)^2 + (y - 2)^2}.$$

It will be easier to minimize $D^2$ than to attempt to minimize $D$:

$$S = D^2 = (x^2 - 8x + 16) + (y^2 - 4y + 4)$$
$$= x^2 - 8x + y^2 - 4y + 20.$$

Solving for $y$, using the equation of the curve, gives:

$$S = x^2 - 8 \cdot 2^{1/2} x^{1/2} + 20.$$
$$\frac{dS}{dx} = 2x - 4\sqrt{2}x^{-1/2} = 0.$$
$$x = 2; \, y = 4.$$

**4.** Maximize $A = xy$ subject to $2x + 2y = k$: $A = \frac{1}{2}kx - x^2$. $dA/dx = \frac{1}{2}k - 2x = 0$. $x = \frac{1}{4}k$; $y = \frac{1}{4}k$. The shape with maximum area will be a square.

**5.** Minimize $z = 2x + 2y$ subject to $xy = A$ : $z = 2x + 2A/x$. $dz/dx = 2 - 2A/x^2 = 0$. $x = \sqrt{A}$. The shape with minimum perimeter will be a square.

**6.** $dY/dx = P - (dC/dx) = 0$. $P = dC/dx = MC$. To maximize profits the firm must choose $x$ so that price is equal to marginal cost.

**7.** $V = hxy$. $y = V/xh$. (cost) $= C = 3rxy + 2rxh + 2ryh$. $dC/dx = 2rh - 2rV/x^2 = 0$. $x^2 = V/h$. $x = y$.

**8.** $C = 2rxy + 2ryh + rxh + 2rxh$
$= 2rV/h + 2rV/x + 3rxh$.
$dC/dx = -2rV/x^2 + 3rh$. $x = (2V/3h)^{1/2}$; $y = (3V/2h)^{1/2}$.

**9.** $V =$ volume $= \pi r^2 h$.
$S =$ (surface area) $= 2\pi rh + 2\pi r^2$
$= 2V/r + 2\pi r^2$.
$dS/dr = -2V/r^2 + 4\pi r = 0$. $r^3 = V/2\pi$; $r = (V/2\pi)^{1/3}$.

**10.** $y' = 4x^3 - 9x^2 + 5$. $y'' = 12x^2 - 18x = 0$. $x = 0, x = 3/2$. $x = 0$ is a local maximum. $x = 3/2$ is a local minimum. The absolute maximum will occur at plus infinity.

**11.** Let the origin be at the point 20 meters due north of South Beach. (Notice that this point is also 20 meters due west of East Beach.) Let $x$ equal the distance from Mongol's boat to the origin, and $y$ be the distance from Recordis' boat to the origin. Then $y = 2000 - 5t$, and $x = 2000 - 7t$. Let $D$ be the distance between the two boats. $D = \sqrt{x^2 + y^2}$. $D^2 = 2000^2 \times 2 + 74t^2 - 48,000t$. To minimize $D^2$, $t = 48,000/148 = 324.32$. When $t = 324.32$, $D = 465$ cm.

**12.** Let $y$ equal the distance that the weight has moved off the ground, and $x$ be the distance from Mongol to the point on the ground directly under the pulley. $dx/dt = u$. $h^2 + x^2 = (L - h + y)^2$. $dy/dx = x/(L - h + y)$. $dy/dt = ux/(L - h + y)$.

**13.** Let $f$ be the amount of fuel that the car uses per hour. Then the total fuel used on the trip will be $F = Tf$, where $T$ is the total time that the trip takes. $T = D/v$.
$F = (D/v)(10v^2 - 100v + 290)$
$= D(10v - 100 + 290/v)$.
$dF/dv = D(10 - 290/v^2) = 0$. $10 = 290/v^2$. $v = \sqrt{29} = 5.39$.

## Chapter 7

**1.** $(1/3)x^3 + (3/2)x^2 + 5x + C$.

**2.** $(a/3)x^3 + (b/2)x^2 + cx + C$.

**3.** $(9/2)x^2 + 10x + C$.

**4.** $14x + C$.

**5.** $\frac{1}{4}x^4 + x + C$.

**6.** $x^6 + (5/2)x^4 + (3/2)x^2 + C$.

**7.** $x^4 + x^3 + x^2 + x + C$.

**8.** $\frac{1}{12}x^4 + \frac{1}{6}x^3 + \frac{1}{2}x^2 + C$.

**9.** $x^{m+1}/(m+1) + x^{n+1}/(n+1) + C$.

**10.** $x^{m+1} + x^{n+1} + x^{p+1} + C$.

**11.** $(1/101)x^{101} + C$.

**12.** $-\cos\theta + C$.

**13.** $\sin\theta + C$.

**14.** $\tan\theta + C$ (remember what the derivative of the tangent function is).

**15.** $(-\frac{1}{3})x^{-3} - \frac{1}{2}x^{-2} + C$.

**16.** $\frac{1}{3}x^3 - x^{-1} + C$.

**17.** $\operatorname{ctn}\theta + C$.

**18.** $\sec\theta + C$.

**19.** $\csc\theta + C$.

**20.** $\frac{1}{2}\theta - \frac{1}{4}\sin 2\theta + C$.

**21.** Use this identity: $\sin^2\theta = \frac{1}{2}(1 - \cos 2\theta)$. The answer is the same as for exercise 20: $\frac{1}{2}\theta - \frac{1}{4}\sin 2\theta + C$.

**22.** $x \int x^2\, dx = \frac{1}{3}x^4 + Cx$. $\int x^3\, dx = \frac{1}{4}x^4 + C$. The two results are different, showing that a variable such as $x$ cannot be moved across an integral sign.

**23.** $x = 5$. This is an object that is not moving at all, such as Hasselbluff Mountain.

**24.** $x = 4t + 2$. This object moves with a constant velocity, such as an ice skater on a frictionless lake.

**25.** $x = 55t - 265$. If $t$ is measured in hours and $x$ in miles, this object is a driver who leaves home at 8 o'clock in the morning, enters the freeway at mile 175, and drives at a constant speed of 55 miles per hour.

**26.** $x = \frac{1}{2}at^2$. This object moves with a constant acceleration, as the train did when Mongol pushed it.

**27.** $x = \frac{1}{2}at^2 + x_0$. This equation represents the time Mongol started pushing the train at the point where $x = x_0$.

**28.** $x = (1/a)\sin(at) + \frac{1}{2} - (1/a)$. This object is a weight attached to a spring. It keeps moving back and forth.

**29.** $x = \frac{1}{5}t^5 + \frac{4}{5}$. I don't know any objects that move like this.

**30.** Integrate once to get $dx/dt$: $dx/dt = \int a\, dt = at + C$. Now use the initial condition $dx/dt = v_0$ when $t = 0$ to show that $dx/dt = at + v_0$. Now integrate again to find $x$: $x = \int(at + v_0)\, dt$. $x = \frac{1}{2}at^2 + v_0 t + C$. Use the initial condition $x = x_0$ when $t = 0$ to show that $x = \frac{1}{2}at^2 + v_0 t + x_0$. This represents an object that starts at $x_0$ with velocity $v_0$ and then moves with constant acceleration, as Mongol's ball does when he throws it into the air with initial velocity $v_0$ from height $x_0$.

**31.** Integrating once gives $dx/dt = C$. Using the initial condition gives $dx/dt = v_0$. Integrating again gives $x = v_0 t + x_0$. This object is another ice skater.

**32.** $dy = 2x\, dx$.

**33.** $dy = \cos x\, dx$.

**34.** $dy = c\, dx$.

**35.** $dy = \frac{1}{2}x^{-1/2}\, dx$.

**36.** $dy = x(1 + x^2)^{-1/2}\, dx$.

**37.** $du = -2x\, dx$. $y = -\frac{1}{2}\int u^{1/2}\, du = -\frac{1}{3}u^{3/2} + C = -\frac{1}{3}(1 - x^2)^{3/2} + C$.

**38.** $du = 3bx^2\, dx$. $y = (1/3b)\int u^{1/2}\, du = (2/9b)(a + bx^3)^{3/2} + C$.

**39.** $du = 2x\, dx$. $y = \int \frac{1}{2}\sin u\, du = -\frac{1}{2}\cos(x^2) + C$.

**40.** $du = \cos x\, dx$. $y = \int u^3\, du = \frac{1}{4}\sin^4 x + C$.

**41.** Let $u = 5 + 6x^2$. $du = 12x\, dx$. $y = \frac{1}{12}\int u^{-2}\, du$. $y = -\frac{1}{12}(5 + 6x^2)^{-1}$.

**42.** Let $u = a + x^n$. $du = nx^{n-1}\, dx$. $y = (1/n)\int u^{1/2}\, du$. $y = (2/3n)(a + x^n)^{3/2} + C$.

**43.** Let $u = \tan x$. $du = \sec^2 x\, dx$. $y = \int u^3\, du = \frac{1}{4}\tan^4 x + C$.

**44.** Let $u = x^{10}$. $du = 10x^9\, dx$. $y = \int \sin u\, du\, (\frac{1}{10})$. $y = -\frac{1}{10}\cos x^{10} + C$.

**45.** $y = [u^{n+1}/(n + 1)] + C$.

**46.** $y = -\cos \theta$.

**47.** (a) Integrate once: $v = \int -g\, dt = -gt + C$. Since $v = v_0$ when $t = 0$, $v = -gt + v_0$. (b) Integrating again gives $h = \int(-gt + v_0)\, dt$. $h = -\frac{1}{2}gt^2 + v_0 t + C$. $h = 0$ when $t = 0$, so $C = 0$. $h = -\frac{1}{2}gt^2 + v_0 t$.

**48.** $v = -gt$. $h = -\frac{1}{2}gt^2 + 64$.

## Chapter 8

**1.** Area $= \displaystyle\int_{-\pi/2}^{\pi/2} \cos x\, dx = \sin x\, \Big|_{-\pi/2}^{\pi/2} = 1 - (-1) = 2$.

**2.** Area $= \displaystyle\int_0^{\pi} \tfrac{1}{2}(1 - \cos 2x)\, dx = \tfrac{1}{2}x \Big|_0^{\pi} - \tfrac{1}{4}\sin 2x \Big|_0^{\pi} = \pi/2.$

**3.** Area $= \displaystyle\int_0^{\pi/5} 3\sin 5x\, dx = (-3/5)\cos 5x \Big|_0^{\pi/5} = (-3/5)(\cos \pi -$ 
$\cos 0) = 6/5.$

**4.** $\displaystyle\int_{-2}^{2} (2x^2 - 8)\, dx = (2/3)x^3 \Big|_{-2}^{2} - 8x \Big|_{-2}^{2} = 16/3 + 16/3 - 16 - 16 =$
$-21.3.$ The area is 21.3. (Note that the value of the definite integral is negative, because the function is negative everywhere in the interval where the integration is carried out.)

**5.** $(1/3) + (3/2) + 5.$

**6.** $(9/2)x^2 \Big|_{-3}^{5} + 10x \Big|_{-3}^{5} = 152.$

**7.** $(a/3)(d_2{}^3 - d_1{}^3) + (b/2)(d_2{}^2 - d_1{}^2) + c(d_2 - d_1).$

**8.** $(a/3)(d_2{}^3 - d_1{}^3) + (b/2)(d_2{}^2 - d_1{}^2) + c(d_2 - d_1).$

**9.** $14a.$

**10.** $8/3.$

**11.** $(5/6) + (5/2) + (3/2) = 4.83.$

**12.** $4.$

**13.** $\tfrac{1}{12}a^4 + \tfrac{1}{6}a^3 + \tfrac{1}{2}a^2.$

**14.** $\dfrac{d_2{}^{m+1} - d_1{}^{m+1}}{m + 1} + \dfrac{d_2{}^{n+1} - d_1{}^{n+1}}{n + 1}.$

**15.** $3.$

**16.** $2/101.$

**17.** $-\cos(\pi/4) + \cos 0 = 1 - 2^{-1/2} = 0.293.$

**18.** $\sin(\pi/4) - \sin(-\pi/4) = 2/\sqrt{2} = \sqrt{2} = 1.414.$

**19.** $(-1/3)x^{-3} \Big|_1^2 - \tfrac{1}{2}x^{-2} \Big|_1^2 = \tfrac{2}{3}.$

**20.** $\tfrac{1}{3}x^3 \Big|_1^2 - x^{-1} \Big|_1^2 = 17/6.$

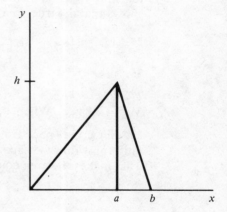

**Figure A–2.**

**21.** The area of the triangle will be the same wherever the triangle is located, so we may as well choose the most convenient location. We can put one vertex of the triangle at the origin, and we can let one side coincide with the $x$ axis. The base of the triangle we can call $b$, and the height we can call $h$ (Figure A–2). We need to solve two integrals to find the total area of the triangle. The first integral represents the area under the line $y = hx/a$, from $x = 0$ to $x = a$. The second integral represents the area under the line $y = hx/(a - b) + bh/(b - a)$, from $x = a$ to $x = b$.

$$A = \int_0^a \frac{hx}{a}\, dx + \int_a^b \left( \frac{hx}{a-b} + \frac{bh}{b-a} \right) dx$$

$$= \frac{h}{a} \left. \tfrac{1}{2}x^2 \right|_0^a + \frac{h}{a-b} \left. \tfrac{1}{2}x^2 \right|_a^b + \frac{bh}{b-a} \left. x \right|_a^b$$

$$= \tfrac{1}{2}ha + \frac{h}{a-b} \tfrac{1}{2}(b^2 - a^2) + bh$$

$$= \tfrac{1}{2}ha + \frac{h}{a-b} \tfrac{1}{2}(b-a)(b+a) + bh$$

$$= \tfrac{1}{2}ha - \tfrac{1}{2}h(b+a) + bh$$
$$= \tfrac{1}{2}ha - \tfrac{1}{2}hb - \tfrac{1}{2}ha + bh$$
$$A = \tfrac{1}{2}bh.$$

**22.** $\displaystyle\int_a^b f(x)\, dx = F(b) - F(a) = -[F(a) - F(b)] = -\int_b^a f(x)\, dx.$

**23.** $y = \left. -\cos\theta \right|_0^{2\pi} = -\cos(2\pi) + \cos 0 = -1 + 1 = 0.$

**24.** (a) $\Delta x = \pi/6.$ $A \approx (\pi/6)(0.5 + 0.8660) = 0.7152.$ (b) $\Delta x = \pi/18.$ $A \approx (\pi/18)\, S = 0.9102.$ (Here $S = \sum_{i=1}^{9} \sin(\pi i/18) = 5.21,$ which is the sum of the first eight elements in Table 8–1.) Evaluating the definite integral shows that the true area is 1.00.

**25.** $y_{\text{av}} = [1/(3-1)] \displaystyle\int_1^3 x^2\, dx = 13/3.$

**26.** $y_{\text{av}} = \tfrac{1}{4}.$

**27.** $V^2_{\text{av}} = \tfrac{1}{2}A^2.$ $V_{\text{rms}} = A\, 2^{-1/2} = 0.707A.$ Since $V_{\text{pp}} = 2A,$ it follows that $V_{\text{pp}} = 2.8 V_{\text{rms}}.$ When $V_{\text{rms}} = 1,$ $V_{\text{pp}} = 2.8.$

**28.** Since the space from $x = 0$ to $x = a$ is divided into $n$ rectangles of uniform width, we can see that $\Delta x = a/n.$ We can also tell that $x_i = ia/n.$ That makes the area equal to:

$$A = \lim_{n\to\infty} \sum_{i=1}^{n} \frac{i^2 a^2}{n^2} \frac{a}{n}.$$

Since $a^3/n^3$ is constant, we can pull it across the summation sign:

$$A = \lim_{n\to\infty} \frac{a^3}{n^3} \sum_{i=1}^{n} i^2.$$

Now we use the formula:

$$A = a^3 \lim_{n\to\infty} \frac{1}{n^3} \frac{n}{6} (2n^2 + 3n + 1)$$

$$= a^3 \lim_{n\to\infty} \left( \frac{1}{3} + \frac{1}{2n} + \frac{1}{6n^2} \right)$$

$$= \tfrac{1}{3}a^3.$$

(This result is the same as the result of the definite integral.)

**1.** $dn/n = 3dt$. $\ln n = 3t + C$. $n = 10$ when $t = 0$, so $C = \ln 10$. $t = \frac{1}{3}$ $\ln(n/10)$. $n = 1000$ when $t = \frac{1}{3} \ln 100 = 1.5$ hours.

**2.** $y = 2 \ln x$. $y' = 2/x$.

**3.** $y' = 1/x$. (Notice that this function is defined only when $x < 0$!)

**4.** $y' = (\sin x)^{-1} \cos x = \text{ctn } x$.

**5.** $y' = (\tan x)^{-1} \sec^2 x$.

**6.** $y = \frac{1}{2} \ln(x^2 + 4)$. $y' = \frac{1}{2}(x^2 + 4)^{-1}(2x) = x/(x^2 + 4)$.

**7.** $y' = (ax^2 + bx + c)^{-1}(2ax + b)$.

**8.** $y = \ln x + \frac{1}{2} \ln(x + 1)$. $y' = x^{-1} + \frac{1}{2}(x + 1)^{-1}$.

**9.**
$$y' = \lim_{\Delta x \to 0} \frac{\log(x + \Delta x) - \log x}{\Delta x} = \lim_{\Delta x \to 0} \frac{\log[(x + \Delta x)/x]}{\Delta x}$$
$$= \lim_{\Delta x \to 0} (x^{-1})(x/\Delta x) \log(1 + \Delta x/x)$$
$$= x^{-1} \log \lim_{\Delta x \to 0} (1 + \Delta x/x)^{x/\Delta x}$$
$$= x^{-1} \log e$$
$y' = 0.434x^{-1}$. (Notice that in this problem $\log x$ has been taken to mean $\log_{10} x$.)

**10.** $y' = x^{-2}(1 - \ln x)$. This curve will have a horizontal tangent where $x = e$. $y'' = x^{-3}(-3 + 2 \ln x)$. This curve will have a point of inflection where $x = e^{3/2} = 4.48$.

**11.** The logarithm of 2 is equal to the area under the curve $y = x^{-1}$ from $x = 1$ to $x = 2$. The width of each of the 10 rectangles is 0.1, and the height is given by the second column of Table 9–3. Adding together the areas of all the rectangles gives the approximate value $\ln 2 = 0.6688$. The best decimal approximation is $\ln 2 = 0.6931$.

**12.**
$$\ln(a/b) = \int_1^{a/b} t^{-1} \, dt = \int_1^a t^{-1} \, dt + \int_a^{a/b} t^{-1} \, dt$$
$$= \ln a + I_2.$$

$I_2 = \int_a^{a/b} t^{-1} \, dt$. Let $u = bt/a$. $du = (b/a) \, dt$.

$$I_2 = \int_b^1 \frac{ab \, du}{ab \, u} = -\int_1^b u^{-1} \, du = -\ln b.$$

$$\ln\left(\frac{a}{b}\right) = \ln a - \ln b.$$

**13.** $\ln(x^n) = \int_1^{x^n} t^{-1} \, dt$. Let $u = t^{1/n}$. $du = (1/n)t^{(1-n)/n} \, dt$.

$$\ln(x^n) = \int_1^x u^{-n} n u^{n-1} \, du = n \int_1^x u^{-1} \, du = n \ln x.$$

**14.** $y = \ln|x + 4| + C.$

**15.** $y = \frac{1}{3}\ln|3x + 4| + C.$

**16.** $y = \int(x + 1)^{-2}\,dx = -(x + 1)^{-1} + C.$

**17.** $y = (1/a)\ln|ax + b| + C.$

**18.** $y = -(ax + b)^{-1}(1/a).$

**19.** Let $u = 3x^2 + 6.$ $du = 6x\,dx.$ $y = \frac{1}{6}\int u^{-1}\,du.$ $y = \frac{1}{6}\ln|3x^2 + 6| + C.$

**20.** Let $u = x^2 + 4x + 6.$ $du = (2x + 4)\,dx.$ $y = \int u^{-1}\,du.$ $y = \ln|x^2 + 4x + 6| + C.$

**21.** Let $u = \cos\theta.$ $du = -\sin\theta\,d\theta.$ $y = -\int u^{-1}\,du = -\ln|\cos\theta| + C.$

**22.** Let $u = \sin\theta + 4.$ $du = \cos\theta\,d\theta.$ $y = \int u^{-1}\,du = \ln|\sin\theta + 4| + C.$

**23.** See Table A–3. Since $f(0)$ is undefined, it should be marked on the graph with an open circle (Figure A–3). It is clear that $\lim_{x \to 0} f(x)$ is somewhere close to 2.8, and of course the actual value of the limit is the mysterious number $e$. As $x$ goes to infinity, the value of the function goes to 1. Note that the curve is not defined for $x < -1$.

| **Table A–3** | |
|---|---|
| $x$ | $(1 + x)^{1/x}$ |
| 10 | 1.271 |
| 8 | 1.316 |
| 6 | 1.383 |
| 4 | 1.495 |
| 2 | 1.732 |
| 1 | 2.000 |
| 0.5 | 2.250 |
| 0.25 | 2.441 |
| 0.1 | 2.594 |
| −0.1 | 2.868 |
| −0.25 | 3.160 |
| −0.5 | 4.000 |
| −0.6 | 4.605 |
| −0.7 | 5.584 |
| −0.8 | 7.477 |
| −0.9 | 12.915 |

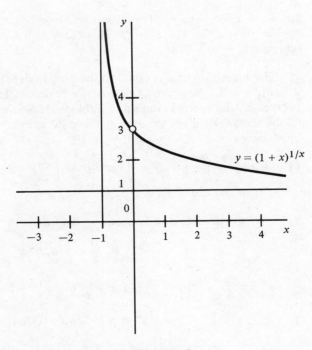

**Figure A–3.**

**24.** Let $K = PV$. Then the work is given by $W = -\int_{V_1}^{V_2} KV^{-1} \, dV$. $W = K$ $\ln(V_1/V_2)$. If $V_1 > V_2$, the gas is being compressed and positive work by some outside force is required to accomplish this. If $V_1 = 2V_2$, then $W = K \ln 2$.

_____ Chapter 10

**1.** $y' = ae^{ax}$.

**2.** $y' = -2xe^{-x^2}$.

**3.** $y' = 2axe^{ax^2+b}$.

**4.** $y' = mae^{\max}$.

**5.** $y' = (\ln 10)10^x$.

**6.** $y' = (e^{-ax}) \cos x - ae^{-ax} \sin x$.

**7.** $\ln y = x^x \ln a$. $(1/y)(dy/dx) = \ln a(x^x + x^x \ln x)$. $dy/dx = a^{x^x} \ln a(x^x + x^x \ln x)$.

**8.** $\ln y = \frac{1}{2}\ln(x - a) + \frac{1}{2}\ln(x - b)$. $(1/y)(dy/dx) = 1/2(x - a) + 1/2(x - b)$. $dy/dx = (x - a)^{1/2}(x - b)^{1/2}[\frac{1}{2}(x - a)^{-1} + \frac{1}{2}(x - b)^{-1}]$.

**9.** $\ln y = \ln(x + 1) - (3/2)\ln(ax^2 + bx + c)$. $(1/y)(dy/dx) = [1/(x + 1)] - (3/2)(2ax + b)(ax^2 + bx + c)^{-1}$. $dy/dx = (ax^2 + bx + c)^{-3/2} - (3/2)(x + 1)(2ax + b)(ax^2 + bx + c)^{-5/2}$.

**10.** $\ln y = \ln(x - 5) - \ln(x + 2) - \ln(x + 3)$. $y' = y[(x - 5)^{-1} - (x + 2)^{-1} - (x + 3)^{-1}]$.

**11.** $y' = (2\pi)^{-1/2} e^{-[(x - \mu)^2/\sigma^2]} [-2(x - \mu)/\sigma^2]$.

**12.** $y = x^n$. $\ln y = n \ln x$. $(1/y)(dy/dx) = nx^{-1}$. $dy/dx = nx^n x^{-1}$. $dy/dx = nx^{n-1}$.

**13.** $y = uv$. $\ln y = \ln u + \ln v$.

$$\frac{1}{y}\frac{dy}{dx} = \frac{1}{u}\frac{du}{dx} + \frac{1}{v}\frac{dv}{dx}$$

$$\frac{dy}{dx} = uv\left(\frac{1}{u}\frac{du}{dx} + \frac{1}{v}\frac{dv}{dx}\right) = v\frac{du}{dx} + u\frac{dv}{dx}.$$

**14.**
$$\frac{f(x + \Delta x) - f(x)}{\Delta x} = \frac{e^{x+\Delta x} - e^x}{\Delta x} = \frac{e^x e^{\Delta x} - e^x}{\Delta x}$$

$$= \frac{e^x(e^{\Delta x} - 1)}{\Delta x} = \frac{e^x(1 + \Delta x + \Delta x^2/2! + \Delta x^3/3! + \cdots - 1)}{\Delta x}.$$

$$\frac{dy}{dx} = \lim_{\Delta x \to 0} e^x(1 + \frac{\Delta x}{2!} + \frac{\Delta x^2}{3!} + \cdots)$$
$$= e^x.$$

**15.** $(1/a)e^{ax} + C.$

**16.** $(1/a)e^{ax+b} + C.$

**17.** Let $u = x^2.$ $du = 2x\ dx.$ $y = \int \frac{1}{2}e^u\ du = \frac{1}{2}e^{x^2} + C.$

**18.** Let $u = e^x + 5.$ $du = e^x\ dx.$ $y = \int u^{-1}\ du = \ln|e^x + 5| + C.$

**19.** Let $u = \ln x.$ $du = x^{-1}\ dx.$ $y = \int u\ du = \frac{1}{2}(\ln|x|)^2 + C.$

**20.** $y' = \cos x - (\cos x - x \sin x) = x \sin x.$

**21.** 19.1.

**22.** $y' = (x^2)(-e^{-x}) + (e^{-x})(2x) = xe^{-x}(2 - x).$ This curve has horizontal tangents where $x = 0$ or $x = 2.$

**23.** (a) Let $u = \sin x.$ $dv = \cos x\ dx.$ $du = \cos x\ dx.$ $v = \sin x.$ $y = \sin^2 x - \int \sin x \cos x\ dx = \sin^2 x - y + C.$ $2y = \sin^2 x + C.$ $y = \frac{1}{2}\sin^2 x + C.$ $y = \frac{1}{4}(1 - \cos 2x) + C = -\frac{1}{4}\cos 2x + C.$ (Note that we can consider the $\frac{1}{4}$ term to be part of the constant.) (b) $u = \sin x.$ $du = \cos x\ dx.$ $y = \int u\ du = \frac{1}{2}u^2 + C.$ $y = \frac{1}{2}\sin^2 x + C = -\frac{1}{4}\cos 2x + C.$ (c) $y = \frac{1}{2}\int \sin 2x\ dx = -\frac{1}{4}\cos 2x + C.$ (Remember that $\sin 2x = 2 \sin x \cos x.$) The trigonometric identity method seems to be easiest in this case.

**24.** $3t = \ln(n/10).$ $e^{3t} = n/10.$ $n = 10e^{3t}.$ When $t = 5,$ the number of bacteria will be $n = 10e^{15} = 3.3 \times 10^7.$

**25.** A problem of this type, where two different types of functions are multiplied together, calls for integration by parts. Let $u = x \ln x.$ $dv = dx.$ $du = (1 + \ln x)\ dx.$ $v = x.$ $y = x^2 \ln x - \int x\ dx - \int x \ln x\ dx.$ Our first reaction on seeing that last integral was to become discouraged, since it is no simpler than the original integral. In fact, it is exactly the same as the original integral. But that means we can substitute: $y = \int x \ln x\ dx.$ Then $y = x^2 \ln x - \frac{1}{2}x^2 - y.$ $2y = x^2 \ln x - \frac{1}{2}x^2.$ $y = \frac{1}{2}x^2 \ln x - \frac{1}{4}x^2 + C.$ (Remember this trick, because it can be used often.)

**26.** $y' = \frac{1}{2}x^2(1/x) + x \ln x - \frac{1}{2}x = x \ln x.$

**27.** Using integration by parts: Let $u = x.$ $dv = e^x\ dx.$ $du = dx.$ $v = e^x.$ $y = xe^x - \int e^x\ dx = xe^x - e^x + C.$

**28.** $y' = xe^x + e^x - e^x = xe^x.$

**29.** Let $u = x^2.$ $dv = e^x\ dx.$ $du = 2x\ dx.$ $v = e^x.$ $y = x^2e^x - 2\int xe^x\ dx.$ Now we have to apply integration by parts again, or we can use the results from exercise 27. $y = x^2e^x - 2xe^x + 2e^x + C.$

**30.** $y' = x^2e^x + 2xe^x - 2xe^x - 2e^x + 2e^x = x^2e^x.$

**31.** Let $u = x^2 \ln x.$ $dv = dx.$ $du = x^2x^{-1} + 2x \ln x.$ $v = x.$ $y = x^3 \ln x - \int(x^2 + 2x^2 \ln x)\ dx.$ $y = x^3 \ln x - \frac{1}{3}x^3 - 2y.$ (Here we used the same trick as in exercise 25.) $y = \frac{1}{3}x^3 \ln x - \frac{1}{9}x^3 + C.$

**32.** Let $u = x$. $dv = (1 + x)^{1/2}\, dx$. $du = dx$. $v = \frac{2}{3}(1 + x)^{3/2}$. $y = (2x/3)(1 + x)^{3/2} - \frac{2}{3}\int(1 + x)^{3/2}\, dx$. $y = (2x/3)(1 + x)^{3/2} - (\frac{2}{3})(\frac{2}{5})(1 + x)^{5/2} + C$.

**33.** Let $u = x^2$. $dv = \sin x\, dx$. $du = 2x\, dx$. $v = -\cos x$. $y = -x^2 \cos x + 2\int x \cos x\, dx$. Use integration by parts again: $u = x$. $dv = \cos x\, dx$. $du = dx$. $v = \sin x$. $y = -x^2 \cos x + 2(x \sin x - \int \sin x\, dx)$. $y = -x^2 \cos x + 2x \sin x + 2 \cos x + C$.

**34.** Let $u = e^x$. $dv = \cos x\, dx$. $du = e^x\, dx$. $v = \sin x$. $y = e^x \sin x - \int e^x \sin x\, dx$. Use integration by parts again: $u = e^x$. $dv = \sin x\, dx$. $du = e^x\, dx$. $v = -\cos x$. $y = e^x \sin x + e^x \cos x - \int \cos x e^x\, dx$. Now use the exercise 25 trick again: $y = e^x \sin x + e^x \cos x - y$. $y = \frac{1}{2}e^x(\sin x + \cos x)$.

**35.** Let $u = \sin^{m-1} x$. $dv = \sin x\, dx$. $du = (m - 1)\sin^{m-2} x \cos x\, dx$. $v = -\cos x$.

$$y = -\sin^{m-1} x \cos x + (m - 1) \int \cos^2 x \sin^{m-2} x\, dx$$
$$= -\sin^{m-1} x \cos x + (m - 1) \int(1 - \sin^2 x) \sin^{m-2} x\, dx$$
$$= -\sin^{m-1} x \cos x + (m - 1) \int \sin^{m-2} x\, dx - (m - 1) \int \sin^m x\, dx$$
$$y = -\sin^{m-1} x \cos x + (m - 1) \int \sin^{m-2} x\, dx - (m - 1)y$$
$$my = -\sin^{m-1} x \cos x + (m - 1) \int \sin^{m-2} x\, dx$$
$$y = \int \sin^m x\, dx = -\frac{\sin^{m-1} x \cos x}{m} + \frac{m - 1}{m} \int \sin^{m-2} x\, dx$$

----

## Chapter 11

**1.** Let $x = \tan \theta$. $dx = \sec^2 \theta\, d\theta$. $\theta = \arctan x$.

$$y = \int \frac{\sec^2 \theta\, d\theta}{\tan^2 \theta \sec \theta} = \int \frac{\cos \theta}{\sin^2 \theta}\, d\theta = \int \operatorname{ctn} \theta \csc \theta\, d\theta$$
$$= -(\sin \theta)^{-1} + C. \quad \sin \theta = x/\sqrt{1 + x^2}.$$
$$= -\frac{\sqrt{1 + x^2}}{x} + C.$$

**2.** Let $x = \sec \theta$. $dx = \sec \theta \tan \theta\, d\theta$. $\theta = \operatorname{arcsec} x$.

$$y = \int \frac{\sec \theta \tan \theta\, d\theta}{\sec \theta \tan \theta} = \int d\theta = \theta + C$$
$$= \operatorname{arcsec} x + C.$$

**3.** Let $u = e^x$. $du = e^x\, dx$.

$$y = \int \frac{1}{1 + u^2}\, du = \arctan u + C$$
$$= \arctan e^x + C.$$

**4.** Let $x = (a/b) \sin \theta$. $dx = (a/b) \cos \theta \, d\theta$. $\theta = \arcsin(bx/a)$.

$$y = \int_0^{\pi/2} \sqrt{a^2 - a^2 \sin^2 \theta} \, (a/b) \cos \theta \, d\theta = (a^2/b) \int_0^{\pi/2} \cos^2 \theta \, d\theta$$

$$= \frac{\pi a^2}{4b}.$$

**5.** $y_1 = (-\tfrac{1}{3})(1 - x^2)^{3/2} + C$. To evaluate $y_2$, use the substitution $x = \sin \theta$. $dx = \cos \theta \, d\theta$. $\theta = \arcsin x$. $y_2/x = \int \cos^2 \theta \, d\theta = \tfrac{1}{2}\theta + \tfrac{1}{4}\sin 2\theta + C = \tfrac{1}{2}\arcsin x + \tfrac{1}{2}\sin \theta \cos \theta + C$. We know that $\sin \theta = x$, and $\cos \theta = \sqrt{1 - \sin^2 \theta} = \sqrt{1 - x^2}$. Therefore $y_2 = \tfrac{1}{2}x \arcsin x + \tfrac{1}{2}x^2\sqrt{1 - x^2} + C$.

**6.** To find the area of one quarter of the circle, set up the integral: $A = \int_0^r \sqrt{r^2 - x^2} \, dx$. Let $x = r \sin \theta$. $dx = r \cos \theta \, d\theta$. $\theta = \arcsin(x/r)$. $A = \int_0^{\pi/2} r^2 \cos^2 \theta \, d\theta = \pi r^2/4$. (You can use the result of exercise 4, with $a = r$ and $b = 1$.) The area of the entire circle will therefore be $\pi r^2$.

**7.** The area will be given by the definite integral $A = 2b \int_{a/2}^a \sqrt{1 - x^2/a^2} \, dx$. Make the substitution $x = a \sin \theta$ and the integral becomes:

$$A = 2ab \int_{\pi/6}^{\pi/2} \cos^2 \theta \, d\theta = 2ab\left(\frac{\pi}{6} - \frac{\sqrt{3}}{8}\right) = 0.61ab.$$

**8.** Make this substitution: $x = a \sin \theta$. $z = 4ab \int \cos^2 \theta \, d\theta$.
$z = 2ab \int (1 + \cos 2\theta) \, d\theta$
$\quad = 2ab(\theta + \tfrac{1}{2}\sin 2\theta)$

$$= 2ab\left[\arcsin\left(\frac{x}{a}\right) + \left(\frac{x}{a}\right)\sqrt{1 - x^2/a^2}\right].$$

$$\frac{dz}{dx} = 2ab\left[\frac{1}{a}\left(1 - \frac{x^2}{a^2}\right)^{-1/2} - \frac{x^2}{a^3}\left(1 - \frac{x^2}{a^2}\right)^{-1/2} + \frac{1}{a} \times \sqrt{1 - x^2/a^2}\right]$$

$$= 2b\left(1 - \frac{x^2}{a^2}\right)^{-1/2} - 2bx^2a^{-2}\left(1 - \frac{x^2}{a^2}\right)^{-1/2} + 2b\left(1 - \frac{x^2}{a^2}\right)^{1/2}$$

$$= 2b\left(1 - \frac{x^2}{a^2}\right)^{-1/2}\left(1 - \frac{x^2}{a^2}\right) + 2b\left(1 - \frac{x^2}{a^2}\right)^{1/2}$$

$$\frac{dz}{dx} = 2b\left(1 - \frac{x^2}{a^2}\right)^{1/2} + 2b\left(1 - \frac{x^2}{a^2}\right)^{1/2} = 4b\sqrt{1 - x^2/a^2}.$$

**9.** Let $u = x/a$. $dx = a \, du$. $y = a \int \frac{1}{u^2 + 1} \, du$. $y = a \arctan(x/a) + C$.

**10.** $y = \frac{1}{a^2} \int \frac{1}{(x/a)^2 + 1} \, dx = (1/a) \arctan(x/a) + C$. (Use the result from exercise 9.)

**11.** $y = \frac{1}{5}\arctan(x/5) + C.$ (Use the result from exercise 10.)

**12.** $y = \displaystyle\int \frac{1}{u^2 + 9}\, du = \frac{1}{3}\arctan[(x + 2)/3] + C.$

**13.** $y = \displaystyle\int \frac{1}{u^2 + c^2 - b^2}\, du.$ Define $d$ so that $d^2 = c^2 - b^2.$ $y = (1/d)$ arctan$[(x + b)/d] + C.$

**14.** Let $u = x + \frac{1}{2}.$ $y = \displaystyle\int \frac{1}{u^2 + 19/4}\, du.$ Let $d = \sqrt{19/4}.$ $y = (1/d)$ arctan$[(x + \frac{1}{2})/d] + C.$

**15.** (a) $x = \sin\theta.$ $dx = \cos\theta\, d\theta.$ $\theta = \arcsin x.$ $y = \int \cos^2\theta\, d\theta.$ $y = \frac{1}{2}\theta + \frac{1}{4}\sin 2\theta + C = \frac{1}{2}\arcsin x + \frac{1}{2}x\sqrt{1 - x^2} + C.$ (b) $x = \cos\theta.$ $dx = -\sin\theta\, d\theta.$ $\theta = \arccos x.$ $y = -\int \sin^2\theta\, d\theta = -\frac{1}{2}\theta + \frac{1}{4}\sin 2\theta = -\frac{1}{2}\arccos x + \frac{1}{2}x(1 - x^2)^{1/2} + C.$ You can decide for yourself which method is easier.

**16.** There is one last hope for an integral that seems to defy any other means of solution: integration by parts. Let $u = \arcsin x.$ $dv = dx.$ $du = (1 - x^2)^{-1/2}\, dx.$ $v = x.$ $y = x\arcsin x - \displaystyle\int \frac{x}{\sqrt{1 - x^2}}\, dx.$ This integral can be solved with a regular substitution: Let $u = 1 - x^2.$ $y = x\arcsin x + \sqrt{1 - x^2} + C.$ Differentiating gives:
$$y' = x(1 - x^2)^{-1/2} + \arcsin x + \frac{1}{2}(1 - x^2)^{-1/2}(-2x) = \arcsin x.$$

**17.** Using integration by parts: Let $u = \arctan x.$ $dv = dx.$ $du = [1/(1 + x^2)]\, dx.$ $v = x.$
$$y = x\arctan x - \int \frac{x}{1 + x^2}\, dx.$$
Let $u = 1 + x^2.$ $y = x\arctan x - \frac{1}{2}\ln(1 + x^2) + C.$
$$\frac{dy}{dx} = \arctan x + \frac{x}{1 + x^2} - \frac{1}{2}(1 + x^2)^{-1}(2x) = \arctan x.$$

---

## Chapter 12

**1.** $\dfrac{x + 4}{(x - 3)(x - 2)} = \dfrac{A}{x - 3} + \dfrac{B}{x - 2} = \dfrac{A(x - 2) + B(x - 3)}{(x - 3)(x - 2)}.$
$x + 4 = x(A + B) - 2A - 3B.$ $1 = A + B.$ $4 = -2A - 3B.$ $A = 7,$ $B = -6.$
$$\frac{x + 4}{(x - 3)(x - 2)} = \frac{7}{x - 3} - \frac{6}{x - 2}.$$

**2.** $3x = Ax - \frac{1}{2}A + Bx + \frac{1}{2}B.$
$$\frac{3x}{(x + \frac{1}{2})(x - \frac{1}{2})} = \frac{3}{2x + 1} + \frac{3}{2x - 1}.$$

**3.** $2x - 4 = A(4x + 3) + B(2x - 2).$
$$\frac{2x - 4}{(2x - 2)(4x + 3)} = -\frac{2}{7(2x - 2)} + \frac{11}{7(4x + 3)}.$$

**4.** $1 = A(x + 3) + Bx.$

$$\frac{1}{x(x + 3)} = \frac{1}{3x} - \frac{1}{3(x + 3)}.$$

**5.** $\dfrac{x^2 + x + 1}{(x + 1)^3} = \dfrac{A}{x + 1} + \dfrac{B}{(x + 1)^2} + \dfrac{C}{(x + 1)^3}.$

$$x^2 + x + 1 = A(x + 1)^2 + B(x + 1) + C.$$

The three-equation system becomes: $A = 1.\ 2A + B = 1.\ A + B + C = 1.$
The result is:

$$\frac{x^2 + x + 1}{x^3 + 3x^2 + 3x + 1} = \frac{1}{x + 1} - \frac{1}{(x + 1)^2} + \frac{1}{(x + 1)^3}.$$

**6.** $\dfrac{(7.1)(9 - 0.4z) - 6.4(10 - 0.6z)}{(10 - 0.6z)(9 - 0.4z)} = \dfrac{63.9 - 2.84z - 64 + 3.84z}{(10 - 0.6z)(9 - 0.4z)}$

$$\simeq \frac{z - 0.1}{(10 - 0.6z)(9 - 0.4z)}.$$

**7.** $\dfrac{2 + 2x + 2 - 2x}{2(1 - x)2(1 + x)} = \dfrac{4}{4(1 - x)(1 + x)} = \dfrac{1}{(1 - x)(1 + x)}.$

**8.** $\dfrac{2(x^2 + 2x + 2) + (x - 1)(x - 1)}{(x - 1)(x^2 + 2x + 2)} = \dfrac{3x^2 + 2x + 5}{(x - 1)(x^2 + 2x + 2)}.$

**9.** $y = \displaystyle\int \frac{2x - 7}{(x - 4)(x - 3)}\, dx.$ Partial fractions are called for in a case like this.

$$\frac{2x - 7}{(x - 4)(x - 3)} = \frac{A}{(x - 4)} + \frac{B}{(x - 3)}.$$

$2x - 7 = Ax - 3A + Bx - 4B.\ A = 1.\ B = 1.$

$$y = \int [(x - 4)^{-1} + (x - 3)^{-1}]\, dx = \ln|x - 4| + \ln|x - 3| + C.$$

**10.** $\dfrac{2x + 9}{(\frac{1}{2}x + 5)(x - 1)} = \dfrac{A}{\frac{1}{2}x + 5} + \dfrac{B}{x - 1}.$

$B = 2.\ A = 1.\ y = 2\ln\left|\frac{1}{2}x + 5\right| + 2\ln|x - 1| + C.$

**11.** $y = (-2 - 3\sqrt{3})\ln|x - \sqrt{3}| + (3 + 2\sqrt{3})\ln|x - 1|.$

**12.** $y = \frac{1}{2}\ln|x - 1| + \frac{1}{2}\ln|x + 1| + C.$

**13.** $y = \frac{1}{2}\ln(1 + x) - \frac{1}{2}\ln(1 - x).$

$$y' = \frac{1}{2(1 + x)} + \frac{1}{2(1 - x)} = \frac{1}{(x - 1)(x + 1)}.$$

**14.** $y' = (\sec x + \tan x)^{-1}(\sec x \tan x + \sec^2 x)$

$\qquad = (\sec x + \tan x)^{-1}(\sec x + \tan x)\sec x = \sec x.$

**15.** $y' = -(\cos x)^{-1}(-\sin x) = \tan x.$

**16.** $y' = \frac{1}{2}(x^2 + 2x + 2)^{-1}(2x + 2) - \dfrac{2}{1 + (x + 1)^2} + \dfrac{2}{x - 1}$

$\qquad = \dfrac{x + 1}{x^2 + 2x + 2} - \dfrac{2}{x^2 + 2x + 2} + \dfrac{2}{x - 1}$

$\qquad = \dfrac{2}{x - 1} + \dfrac{x - 1}{x^2 + 2x + 2}.$

**17.** This integral requires integration by parts. It is hard to tell at first glance what the best parts are, but after some trial and error it turns out that the best strategy is to set $u = \sec x$. $dv = \sec^2 x\, dx$. $du = \sec x \tan x\, dx$. $v = \tan x$. $y = \sec x \tan x - \int \tan^2 x \sec x\, dx$.
Using the secant-tangent identity gives:
$$y = \sec x \tan x - \int \sec^3 x\, dx + \int \sec x\, dx$$
$$= \sec x \tan x - y + \ln\left|\sec x + \tan x\right| + C$$
$$= \tfrac{1}{2} \sec x \tan x + \tfrac{1}{2} \ln\left|\sec x + \tan x\right| + C.$$
Differentiating, we obtain:
$$y' = \tfrac{1}{2}(\sec x \sec^2 x + \sec x \tan^2 x) + \tfrac{1}{2} \sec x$$
$$= \tfrac{1}{2}(\sec^3 x + \sec^3 x - \sec x + \sec x)$$
$$= \sec^3 x.$$

**18.** The area is given by $A = 2\displaystyle\int_a^{2a} b\,\sqrt{x^2/a^2 - 1}\, dx$. Make the substitution $x = a \sec \theta$. The final result is $2.14ab$.

**19.** Let $x = \tan \theta$. $y = \int \sec^3 \theta\, d\theta$. Use the result of exercise 17:
$$y = \tfrac{1}{2} \sec \theta \tan \theta + \tfrac{1}{2} \ln\left|\sec \theta + \tan \theta\right| + C$$
$$= \tfrac{1}{2} x\sqrt{1 + x^2} + \tfrac{1}{2} \ln\left|x + \sqrt{1 + x^2}\right| + C.$$

**20.** Let $x = \tan \theta$. $y = \int \sec \theta\, d\theta$. $y = \ln\left|\sec \theta + \tan \theta\right| + C$. $y = \ln\left|x + \sqrt{1 + x^2}\right| + C.$

**21.** Let $x = \sin \theta$.
$$y = \int \cos^2 \theta\,(\sin \theta)^{-1}\, d\theta = \int(\sin \theta)^{-1}\, d\theta - \int \sin \theta\, d\theta$$
$$= \cos \theta - \ln\left|\csc \theta + \operatorname{ctn} \theta\right| + C$$
$$= \sqrt{1 - x^2} - \ln\left|x^{-1} + (1 - x^2)^{1/2}\, x^{-1}\right| + C.$$

**22.** Let $x = \tan \theta$.
$$y = \int(\sin \theta)^{-1}\, d\theta = -\ln\left|\csc \theta + \operatorname{ctn} \theta\right| + C$$
$$= -\ln\left|x^{-1} + (1 - x^2)^{1/2}\, x^{-1}\right| + C.$$

**23.** Let $x = \tan \theta$.
$$y = \int \tan^2 \theta \sec \theta\, d\theta = \int \sec^3 \theta\, d\theta - \int \sec \theta\, d\theta$$
$$= \tfrac{1}{2} \sec \theta \tan \theta - \tfrac{1}{2} \ln\left|\sec \theta + \tan \theta\right| + C$$
$$= \tfrac{1}{2}x\sqrt{x^2 + 1} - \tfrac{1}{2} \ln\left|x + \sqrt{x^2 + 1}\right| + C.$$

**24.** These problems can be done by using algebraic division or synthetic division. Using algebraic division:

$$
\begin{array}{r}
x^2 + \phantom{0}x \phantom{+0000000} \\
2x^2 + 3x + 4)\overline{\phantom{)}2x^4 + 5x^3 + 7x^2 + 5x} \\
\underline{-2x^4 + 3x^3 + 4x^2\phantom{+00000}} \\
2x^3 + 3x^2 + 5x \\
\underline{-2x^3 + 3x^2 + 4x} \\
x
\end{array}
$$

The remainder is $x$, so the final result is

$$x^2 + x + \frac{x}{2x^2 + 3x + 4}.$$

**25.** $5x + [1/(3x^3 - 3)]$.

**26.** $5x^2 + 4x + [1/(x^2 - 1)]$.

**27.** $3x^2 + 2x + 5 + [(4x - 1)/(x^2 - x - 12)]$.

**28.** (a) When $b^2 - 4ac$ is positive, the integral is of the same general family as $\int [1/(x^2 - 1)]\, dx$. The integral can be simplified with the substitution $u = x + \frac{1}{2}b/a$. ($dx = du$.) The result:

$$y = \int \frac{1}{au^2 - \frac{1}{4}D^2/a}.$$

$$du = \frac{4a}{D^2} \int \frac{1}{[(2au)^2/D^2] - 1}\, du.$$

Now let $v = 2au/D$. $y = -\dfrac{2}{D} \int \dfrac{1}{1 - v^2}\, dv$.

This integral can be solved with partial fractions, as was done in the chapter.

$$y = -\frac{2}{D} \int \left[ \frac{1}{2(1 - v)} + \frac{1}{2(1 + v)} \right] dv$$

$$= \frac{1}{D} \ln \left| \frac{v - 1}{v + 1} \right|.$$

Solving for $v$ in terms of $x$ yields $v = (2ax + b)/D$. So the final result is

$$y = \frac{1}{D} \ln \left| \frac{2ax + b - D}{2ax + b + D} \right| + C.$$

(b) When $b^2 - 4ac$ is negative, the integral is of the same general family as $\int [1/(1 + x^2)]\, dx$. Start by making the same substitution: $v = (2ax + b)/D$. After some algebra the integral turns out to be equal to $(2/D) \int [1/(1 + v^2)]\, dv$. That integral is instantly recognizable, so the final answer is

$$y = \frac{2}{D} \arctan\left( \frac{2ax + b}{D} \right) + C.$$

You can see how a table of integrals would spare you from repeatedly performing some of these calculations.

**29.** $b^2 - 4ac = -4$. Use exercise 28, case (b). $y = \arctan(x + 1) + C$.

**30.** $b^2 - 4ac = 13$. Case (a): $y = 13^{-1/2} \ln \left| \dfrac{2x - 3 - \sqrt{13}}{2x - 3 + \sqrt{13}} \right| + C$.

**31.** $b^2 - 4ac = -3$. Case (b): $y = (2/\sqrt{3}) \arctan[(2x + 1)/\sqrt{3}] + C$.

**32.** $b^2 - 4ac = -3$. Case (b): $y = (2/\sqrt{3}) \arctan[(2x - 1)/\sqrt{3}] + C$.

**33.** Factor the denominator: $y = \int \dfrac{1}{(x - 1)(x^2 + x + 1)} \, dx$.
Set up the partial fractions:

$$\frac{1}{(x - 1)(x^2 + x + 1)} = \frac{A}{x - 1} + \frac{Bx + C}{x^2 + x + 1}.$$

$1 = Ax^2 + Ax + A + Bx^2 - Bx + Cx - C. \; A + B = 0. \; A - B + C = 0.$

$A - C = 1. \; A = \frac{1}{3}. \; B = -\frac{1}{3}. \; C = -\frac{2}{3}.$

$$y = \tfrac{1}{3} \ln|x - 1| - \int \frac{x/3 + \frac{2}{3}}{x^2 + x + 1} \, dx$$

$$= \tfrac{1}{3} \ln|x - 1| - \tfrac{1}{6} \ln|x^2 + x + 1| - (\tfrac{2}{3} - \tfrac{1}{6}) \left( \frac{2}{\sqrt{3}} \right)$$

$$\times \arctan\left( \frac{2x + 1}{\sqrt{3}} \right) + C.$$

(Use the result from exercise 31.)

**34.** Factor the denominator: $y = \int \dfrac{1}{(x + 1)(x^2 - x + 1)} \, dx$.
After solving the partial fractions:

$$y = \int \frac{1}{3(x + 1)} \, dx + \int \frac{(-\frac{1}{3})x + \frac{2}{3} \, dx}{x^2 - x + 1}$$

$$= \tfrac{1}{3} \ln|x + 1| - \tfrac{1}{6} \ln|x^2 - x + 1| + (\tfrac{2}{3} + \tfrac{1}{6}) \left( \frac{2}{\sqrt{3}} \right)$$

$$\times \arctan\left( \frac{2x - 1}{\sqrt{3}} \right) + C.$$

**35.** $b^2 - 4ac = 4$. Case (a): $y = \tfrac{1}{2} \ln \left| \dfrac{4x}{4x + 4} \right| + C$.

**36.** $b^2 - 4ac = 13$. Case (a): $y = 13^{-1/2} \ln \left| \dfrac{6x + 5 - \sqrt{13}}{6x + 5 + \sqrt{13}} \right| + C$.

**37.** Factor the denominator: $y = \int \dfrac{x}{(x^2 + 1)(x - 1)(x + 1)} \, dx$.
After solving the partial fractions the result is:

$$y = \int \left[ \frac{-x}{2(x^2 + 1)} + \frac{1}{4(x - 1)} + \frac{1}{4(x + 1)} \right] dx$$

$$= -\tfrac{1}{4} \ln|x^2 + 1| + \tfrac{1}{4} \ln|x - 1| + \tfrac{1}{4} \ln|x + 1| + C.$$

**38.** After solving for the partial fractions:

$$y = \int \left[ \frac{1}{2(x - 1)} - \frac{x + 1}{2(x^2 + 1)} \right] dx$$

$$= \tfrac{1}{2} \ln|x - 1| - \tfrac{1}{4} \ln|x^2 + 1| - \tfrac{1}{2} \arctan x + C.$$

## Chapter 13

**1.** $dV = \pi r^2 \, dx$. $r = y$. $y^2 = b^2(1 - x^2/a^2)$. $V = (4/3)\pi ab^2$.

**2.** $dV = A \, dy$. $A = \frac{1}{2}(2x)(\sqrt{3} \, x) = x^2\sqrt{3}$. $V = \int_0^2 x^2\sqrt{3} \, dy = \int_0^2 2y\sqrt{3} \, dy = 4\sqrt{3}$.

**3.** $V = \int_4^8 \pi y \, dy = 24\pi$.

**4.** (a) $V = \int_a^b \pi[f(x)]^2 \, dx$. (b) $V = \int_a^b 2\pi(b - x)f(x) \, dx$. (Use cylindrical shells.) (c) $V = \int_a^b 2\pi(c - x)f(x) \, dx$.

**5.** (a) $V(a) = 0$. (b) $V(x + \Delta x) - V(x) = \pi[f(x)]^2 \, \Delta x$. $[V(x + \Delta x) - V(x)]/\Delta x = \pi[f(x)]^2$. $dV/dx = \pi[f(x)]^2$. $\int dV = \int \pi[f(x)]^2 \, dx$. $V(x) = F(x) + C$. $V(x) = 0$ when $x = a$. Solving for the arbitrary constant gives $C = -F(a)$. Therefore $V(x) = F(x) - F(a)$. The total volume of the solid will be $V = F(b) - F(a)$.

**6.** $V = \int_{r/2}^r \pi(r^2 - y^2) \, dy$. $V = (5/24)\pi r^3$.

**7.** $V = \int_{-r}^r \pi(r^2 - y^2) \, dy$. $V = \pi 2r^3 - \pi(2/3)r^3 = (4/3)\pi r^3$.

**8.** Divide the pyramid into little squares of side $x$. Let $z$ be the distance from the top of the pyramid to each square. Let $h$ be the height of the pyramid, and $s^2$ be the area of the base. $z/x = h/s$ (by similar triangles).

$$V = \int_{-h}^0 x^2 \, dz = \frac{1}{3}s^2h.$$

## Chapter 14

**1.** $1 + (dy/dx)^2 = 1 + 9x/4$. $L = 2.09$.

**2.** $1 + (dy/dx)^2 = 1 + x^2$. $L = \int_0^a \sqrt{1 + x^2} \, dx$. $L = \frac{1}{2}a\sqrt{1 + a^2} + \frac{1}{2} \ln (a + \sqrt{1 + a^2})$. When $a = 2$, $L = 2.96$.

**3.** $1 + (dy/dx)^2 = 1 + \tan^2 x = \sec^2 x$. $L = 0.88$. The length of the curve from $x = 0$ to $x = \pi/2$ is infinity.

**4.** $1 + (dy/dx)^2 = 1 + a^2$. $L = b\sqrt{1 + a^2}$.

**5.** $L = \int_0^a \sqrt{1 + \frac{1}{4}x^{-1}} \, dx$.

**6.** $L = \int_1^a \sqrt{1 + x^{-2}}\, dx.$

**7.** $L = \int_0^\pi \sqrt{1 + \cos^2 x}\, dx.$

**8.** $1 + (dx/dy)^2 = 1 + 1/4y.\; A = 2\pi \int_{3/4}^{15/4} \sqrt{y}\sqrt{1 + 1/4y}\, dy.\; A = \pi \int_{3/4}^{15/4} \sqrt{4y + 1}\, dy = 28\pi/3.$

**9.** $1 + (dy/dx)^2 = 1 + a^2.\; A = 2\pi a\sqrt{1 + a^2} \int_0^b x\, dx.\; A = \pi a b^2 \sqrt{1 + a^2}.$

**10.** $1 + (dy/dx)^2 = 1 + \cos^2 x.\; A = 2\pi \int_0^\pi \sin\, x\, \sqrt{1 + \cos^2 x}\, dx.$ Let $u = \cos x.\; A = \pi[2\sqrt{2} + \ln(1 + \sqrt{2}) - \ln(\sqrt{2} - 1)] = 14.42.$

**11.** $1 + (dy/dx)^2 = 1 + x^4.\; A = 2\pi \int_0^a \tfrac{1}{3} x^3 \sqrt{1 + x^4}\, dx.$ Let $u = 1 + x^4.$ $A = (\pi/6) \int_1^{1+a^4} u^{1/2}\, du.\; A = (\pi/9)[(1 + a^4)^{3/2} - 1].$ When $a = 2$, the surface area is 24.12.

**12.** $1 + (dy/dx)^2 = 1 + e^{2x}.\; A = 2\pi \int_0^a e^x \sqrt{1 + e^{2x}}\, dx.$ Let $u = e^x.\; A = 2\pi \int_1^{e^a} \sqrt{1 + u^2}\, du.\; A = \pi[e^a \sqrt{1 + e^{2a}} + \ln(e^a + \sqrt{1 + e^{2a}}) - \sqrt{2} - \ln(1 + \sqrt{2})].$

**13.** $A = \pi r^2.$

**14.** $dm = \rho\, dv.\; dV = 2xh\, dy.\; y_{com} = \int_0^a y\, dm / \int_0^a dm.\; y_{com} = 3a/5.$

**15.** $dm = \rho\pi x^2\, dy.\; y_{com} = 3b/4.$

**16.** $x_{com} = \dfrac{\int_a^b x[f(x)]^2\, dx}{\int_a^b [f(x)]^2\, dx}.$

**17.** $dm = \rho\, dV.\; \rho = \sqrt{x} + 2.\; dV = A\, dx.$
$$x_{com} = \frac{\int_0^L x\, dm}{\int_0^L dm} = \frac{\int_0^L (x^{3/2} + 2x)\, dx}{\int_0^L (x^{1/2} + 2)\, dx}$$
$$= \frac{0.4 L^{2.5} + L^2}{\tfrac{2}{3} L^{1.5} + 2L}.$$

**18.** $I = \int_0^L x^2 \rho A \; dx = \rho A \; \frac{1}{3}x^3 \Big|_0^L = ML^2/3.$

**19.** $I = 2\pi h \rho R^4/4 = MR^2/2.$

**20.** $dm = 2\pi r \rho (2L) \; dr. \; L = (R^2 - r^2)^{1/2}.$

$I = 4\pi\rho \int_0^R r^3 \sqrt{R^2 - r^2} \; dr.$ (Remember that $R$ is constant and $r$ is a variable.) Let $r = R \sin \theta.$ $I = 4\pi\rho R^5 \int_0^{\pi/2} \sin^3 \theta \cos^2 \theta \; d\theta.$ $I = 4\pi\rho R^5 \int_0^{\pi/2} (\sin \theta \cos^2 \theta - \sin \theta \cos^4 \theta) \; d\theta.$ Let $u = \cos \theta.$ $du = -\sin \theta \; d\theta.$

$$I = 4\pi\rho R^5 \int_1^0 (-u^2 + u^4) \; du$$
$$= 4\pi\rho R^5 (\tfrac{1}{3} - \tfrac{1}{5})$$
$$= \tfrac{4}{3}\pi R^3 \rho R^2 (\tfrac{2}{5})$$
$$= \frac{2MR^2}{5}.$$

**21.** $I = 2\pi\rho \int_0^R x^3 (h - hx/R) \; dx. \; I = 3MR^2/10.$

## Chapter 15

**1.** Nonlinear (because of the $y^2$ term).

**2.** Linear and nonhomogeneous.

**3.** Nonlinear (because of $(dy/dx)^2$).

**4.** Nonlinear (unless $f(y)$ is a linear function of $y$).

**5.** Linear and nonhomogeneous.

**6.** Nonlinear.

**7.** Nonlinear.

**8.** $T(ax_1 + bx_2)$

$$= \frac{d^n (ax_1 + bx_2)}{dt^n} + \cdots + f_1(t) \frac{d}{dt} (ax_1 + bx_2) + f_0(t)(ax_1 + bx_2)$$
$$= a \frac{d^n}{dt^n} x_1 + b \frac{d^n}{dt^n} x_2 + \cdots + af_1(t) \frac{dx_1}{dt}$$
$$\quad + bf_1(t) \frac{dx_2}{dt} + af_0(t)x_1 + bf_0(t)x_2$$
$$= a \left[ \frac{d^n}{dt^n} + \cdots + f_1(t) \frac{d}{dt} + f_0(t) \right] x_1$$

$$+ b\left[\frac{d^n}{dt^n} + \cdots + f_1(t)\frac{d}{dt} + f_0(t)\right]x_2$$

$$T(ax_1 + bx_2) = aTx_1 + bTx_2.$$

**9.** Set up the characteristic equation: $r^2 + r - 6 = 0$. $r = 2, -3$. $x = Ae^{2t} + Be^{-3t}$.

**10.** $r^2 + r + 1 = 0$. $r = \frac{1}{2}(-1 \pm i\sqrt{3})$. $x = e^{-t/2}(A \sin \sqrt{3}t + B \cos \sqrt{3}t)$.

**11.** $r^2 + 9 = 0$. $r = \pm 3i$. $x = A \sin 3t + B \cos 3t$.

**12.** $r^2 + 2r + 1 = 0$. $r = -1$. $x = Ae^{-t} + Bte^{-t}$.

**13.** $r^2 - 4r = 0$. $r = 0, 4$. $x = A + Be^{4t}$.

**14.** In order for the characteristic equation to have one real root, the differential equation must be of the form:

$$\frac{d^2x}{dt^2} + b\frac{dx}{dt} + \frac{b^2x}{4} = 0.$$

When $x = Ae^{rt} + Bte^{rt}$, $dx/dt = (Ar + B)e^{rt} + Brte^{rt}$, and $d^2x/dt^2 = (Ar^2 + 2Br)e^{rt} + Br^2te^{rt}$. (Note that $b = -2r$.) Therefore:

$$\frac{d^2x}{dt^2} + b\frac{dx}{dt} + \frac{b^2x}{4} = (Ar^2 + 2Br)e^{rt} + Br^2te^{rt} - 2(Ar^2 + Br)e^{rt}$$

$$- 2Br^2te^{rt} + r^2Ae^{rt} + Btr^2e^{rt}$$

$$= 0.$$

**15.** $x = 0.658 \sin(2t - 1.4) + e^{-1.5t} A_1 \sin(1.66 t + A_2)$.

**16.** $-0.6484 = 0.658 \sin(-1.4) + A_1 \sin A_2$. $1.884 = 2(0.658) \cos(-1.4) + 1.66A_1 \cos A_2 + 1.5A_1 \sin A_2$. $0 = A_1 \sin A_2$. $1.66 = 1.66A_1 \cos A_2 + 1.5A_1 \sin A_2$. $A_2 = 0$, $A_1 = 1$. $x = 0.658 \sin(2t - 1.4) + e^{-1.5t} \sin(1.66 t)$.

**17.** The equation of motion in this case is: $(d^2x/dt^2) + 4(dx/dt) + 4x = 0$. $r^2 + 4r + 4 = 0$. $r = -2$. $x = Ae^{-2t} + Bte^{-2t}$. Notice that, if the friction is strong enough, the ride will not oscillate at all. This situation is known as *critical damping*.

**18.** The characteristic equation is: $r^2 + 5r + 4 = 0$. $r = -1$, or $r = -4$. $x = Ae^{-t} + Be^{-4t}$. This situation is known as *overdamping*.

# Index